CAMBRIDGE MONOGRAPHS
ON APPLIED AND COMPUTATIONAL
MATHEMATICS

46 Kernel-Based Methods: Milestones and New Trends

The *Cambridge Monographs on Applied and Computational Mathematics* series reflects the crucial role of mathematical and computational techniques in contemporary science. The series publishes expositions on all aspects of applicable and numerical mathematics, with an emphasis on new developments in this fast-moving area of research.

State-of-the-art methods and algorithms as well as modern mathematical descriptions of physical and mechanical ideas are presented in a manner suited to graduate research students and professionals alike. Sound pedagogical presentation is a prerequisite. It is intended that books in the series will serve to inform a new generation of researchers.

A complete list of books in the series can be found at www.cambridge.org/mathematics.

Recent titles include the following:

32. Multivariate approximation, *V. Temlyakov*
33. Mathematical modelling of the human cardiovascular system, *Alfio Quarteroni, Luca Dede', Andrea Manzoni & Christian Vergara*
34. Numerical bifurcation analysis of maps, *Yuri A. Kuznetsov & Hil G.E. Meijer*
35. Operator-adapted wavelets, fast solvers, and numerical homogenization, *Houman Owhadi & Clint Scovel*
36. Spaces of measures and their applications to structured population models, *Christian Düll, Piotr Gwiazda, Anna Marciniak-Czochra & Jakub Skrzeczkowski*
37. Quasi-interpolation, *Martin Buhmann & Janin Jäger*
38. The Christoffel–Darboux kernel for data analysis, *Jean Bernard Lasserre, Edouard Pauwels & Mihai Putinar*
39. Geometry of the phase retrieval problem, *Alexander H. Barnett, Charles L. Epstein, Leslie Greengard & Jeremy Magland*
40. Discrete variational problems with interfaces, *Roberto Alicandro, Andrea Braides, Marco Cicalese & Margherita Solci*
41. Spectral and spectral element methods for fractional ordinary and partial differential equations, *Mohsen Zayernouri, Li-Lian Wang, Jie Shen & George Em Karniadakis*
42. High-accuracy finite difference methods, *Bengt Fornberg*
43. An introduction to immersed boundary methods, *Roberto Verzicco, Marco D. de Tullio & Francesco Viola*
44. Differential equations and variational methods on graphs, *Yves van Gennip & Jeremy Budd*
45. Minimal cubature rules, *Yuan Xu*
46. Kernel-Based Methods: Milestones and New Trends, *Stefano De Marchi, Francesco Marchetii & Emma Perracchione*

Kernel-Based Methods

Milestones and New Trends

STEFANO DE MARCHI
University of Padova

FRANCESCO MARCHETTI
University of Padova

EMMA PERRACCHIONE
Polytechnic of Torino

CAMBRIDGE
UNIVERSITY PRESS

CAMBRIDGE
UNIVERSITY PRESS

Shaftesbury Road, Cambridge CB2 8EA, United Kingdom

One Liberty Plaza, 20th Floor, New York, NY 10006, USA

477 Williamstown Road, Port Melbourne, VIC 3207, Australia

314–321, 3rd Floor, Plot 3, Splendor Forum, Jasola District Centre,
New Delhi – 110025, India

Cambridge University Press is part of Cambridge University Press & Assessment,
a department of the University of Cambridge.

We share the University's mission to contribute to society through the pursuit of
education, learning and research at the highest international levels of excellence.

www.cambridge.org
Information on this title: www.cambridge.org/9781009769129
DOI: 10.1017/9781009769105

When citing this work, please include a reference to the DOI 10.1017/9781009769105

First published 2026

A catalogue record for this publication is available from the British Library

A Cataloging-in-Publication data record for this book is available from the Library of Congress

ISBN 978-1-009-76912-9 Hardback

To my wife, Cristina, and daughter, Silvia, for their uncountable support throughout my career, making my life always a "nice and sweet adventure"

— with love, Stefano

[A mia moglie Cristina e a mia figlia Silvia, per il loro innumerevole supporto durante la mia carriera. Hanno reso la mia vita una "bella e dolce avventura"

— con amore, Stefano]

To my sparring friend, my mum Luisella, so delightfully unlike me and still by my side many years on

— your beast, Emma

[Alla mia amica di battibecchi, la mia mamma Luisella, così deliziosamente diversa da me e ancora al mio fianco molti anni dopo

— la tua bestia, Emma]

To my daughter Aurora, who'll learn that in life, as in math, sometimes the best we can do is to approximate

— your dad Francesco

[A mia figlia Aurora, che imparerà che nella vita, come in matematica, a volte il meglio che possiamo fare è approssimare

— tuo papà Francesco]

"This is an extremely helpful book for researchers who work with kernels in the context of modern applications – including machine learning, image processing, and data science. It provides important new theoretical results as well as numerous practical examples, and supplies the necessary software to open up new kernel-based fields of research."

Robert Schaback, University of Göttingen

"This volume complements the existing textbooks on kernel methods with exciting new topics such as rational kernel approximation, approximation of discontinuous data, and learning with kernels."

Elisabeth Larsson, Uppsala University

"Kernel and radial basis function-based approximations are at the very centre of modern multivariate approximation theory and applications. This book reviews state-of-the-art approaches including excellent summaries of the basic theoretical results on kernel-based methods and their applications; the practical aspects of these tools are particularly emphasized and a great number of very useful tools, including software, is included. This is a book that should be in every application-interested scientist who works with multivariate approximations."

Martin Buhmann, University of Giessen

"This monograph complements existing books on kernels and their applications by focusing on promising recent developments – such as rational kernels, data-driven kernels, variably scaled kernels, persistence kernels – the details of which were previously accessible only in original research publications."

Oleg Davydov, University of Giessen

"This book seamlessly bridges rigorous mathematical theory with practical, hands-on applications. Readers will explore cutting-edge solutions for scattered data problems – with a special focus on variably scaled kernels (VSKs) – alongside imaging challenges and vital machine learning tools like support vector machines (SVMs). To ensure immediate implementation, the text introduces key software packages with direct links to open-source code repositories on GitHub. Designed for students, applied scientists, and practitioners alike, this text bypasses advanced functional analysis to deliver a direct, comprehensive understanding of kernel-based algorithms."

Greg Fasshauer, Colorado School of Mines

Contents

Figures

Tables

Preface

Why This Book

This book was intended as lecture notes used by the authors in some master's and PhD courses taught in the last decade. The manuscript drew inspiration from two seminal books on kernel approximation, namely those by Holger Wendland (2005) and Gregory E. Fasshauer (2007). The former provides a rigorous and elegant treatment of the theoretical foundations. In contrast, the second combines theory with a practical approach and useful MATLAB functions that greatly facilitate understanding of both the mathematical principles and their applications.

The core material of these notes was initially prepared in 2016 for the master's course on approximation theory at the University of Padova by Stefano De Marchi. Later, expanded versions of this manuscript, written by Emma Perracchione, were used during teaching visits to the University of Antwerp, within the framework of the Erasmus Teaching Staff Mobility, and in the Allameh Tabataba'i University of Tehran for the workshops "Meshless Methods: Theory, Algorithms, Applications" (2018) and "Softwares and Meshless Methods and Applications in Finance" (2019).

The discussions and feedback received from students and colleagues on those occasions have been invaluable in refining and extending the material that formed the basis of this book. Therefore, together with Francesco Marchetti, we decided to write the present volume, whose primary goal is to collect the most relevant advances proposed by the authors over the past few years in the field of approximation theory with Kernels.

Indeed, close collaborations among the authors of this manuscript have been established since 2017. In particular, both Francesco Marchetti and Emma Perracchione have been working for several years under the supervision of Stefano De Marchi.

About This Book

Over the past two decades, the literature on kernels and radial basis functions has grown at an impressive rate, reflecting their increasing relevance across a wide range of scientific and engineering disciplines. Those who wish to deepen their knowledge will find an extensive and ever-expanding body of work devoted to this topic.

The popularity of kernels can perhaps be best illustrated through a simple analogy. In many culinary traditions, parsley is used to enhance the flavor and appearance of dishes. Its presence is subtle yet almost ubiquitous. Likewise, kernel-based methods have become a kind of mathematical parsley: A versatile and powerful ingredient that appears in nearly all problems that require an efficient and stable approximation tool. Whether in numerical analysis, data interpolation, machine learning, or even finance, their unifying properties and flexibility have made them indispensable in both theory and practice.

These lectures were designed with the needs of students and young researchers in mind, especially those without a strong background in functional analysis. For this reason, we have chosen, whenever possible, to avoid introducing advanced concepts from that field. This omission is certainly a limitation, yet we hope readers will not be too disappointed and instead appreciate the more direct and intuitive approach we have adopted. The goal is to make the subject accessible while preserving the mathematical rigor necessary to understand its depth. In view of this, the target audience also includes machine learning practitioners who want to deepen their understanding of the mathematical foundations of kernel-based algorithms, as well as applied scientists working with scattered data, imaging problems, or classification tasks.

It should also be emphasized that many essential aspects, extensions, and applications are not covered here, partly because they are already treated thoroughly in other excellent references. Nonetheless, these notes include several recent research topics developed by the authors. The combination of theoretical insight and numerical experimentation is intended to provide a more engaging and balanced learning experience, helping readers connect abstract ideas with practical numerical experiments.

We sincerely hope that this introductory text will stimulate curiosity and inspire students and researchers alike to explore this fascinating mathematical tool further. The theory of kernel methods stands at the crossroads of analysis, computation, and application, offering not only elegant mathematical structures but also practical solutions to real-world problems. If these notes succeed in encouraging readers to look deeper into the field, or even to contribute to its ongoing development, they will have achieved their purpose.

Acknowledgments

We begin by expressing our sincere gratitude to the students who attended the various master's and PhD courses in which parts of these notes were initially conceived and refined. Their thoughtful questions, insightful comments, and genuine curiosity played a crucial role in shaping the evolution of this material. Often, it was through their attentive participation that we discovered which explanations were transparent and which required deeper clarification or a more intuitive presentation. The enthusiasm they brought to each lecture not only enriched the classroom dynamic but helped illuminate the aspects of kernel-based methods that are most intriguing – and sometimes most challenging – for newcomers to the field. The lectures delivered at the University of Padova, the University of Antwerp, and the Allameh Tabataba'i University of Tehran provided particularly stimulating academic environments. Each institution offered a unique context for testing, adapting, and expanding the material. At the University of Padova, the long-standing tradition of numerical analysis and approximation theory provided fertile ground for some of the book's foundational chapters. In Antwerp, our interactions with international students and colleagues broadened our perspective on how kernel methods are perceived and applied across different research communities. The workshops and lectures in Tehran, characterized by a vibrant enthusiasm for meshless methods, offered an invaluable opportunity to explore emerging ideas and to appreciate the growing global interest in kernel-based techniques.

We are deeply grateful to our colleagues and collaborators, whose expertise and enthusiasm have inspired many of the ideas presented here. Over the years, countless discussions have helped to refine the material. Many persons and institutions have contributed, directly or indirectly, to the development of this book. It is a pleasure to acknowledge their support and express our gratitude. While it would be impossible to thank everyone individually, we warmly recognize the collective contribution of the scientific community, whose inspiration, challenges, and guidance have accompanied us throughout this journey.

On a more personal level, we extend heartfelt thanks to our families. Their encouragement, understanding, and patience made it possible to devote the extended time required to bring this manuscript to completion.

Finally, we thank the community of researchers and practitioners who continue to advance the theory and applications of kernel methods. The interactions, exchanges of ideas, and shared enthusiasm for this fascinating area of mathematics are among the primary reasons we remain passionate about this field. We hope that this book, in turn, will assist others in their own explorations and contributions.

We are indebted to the institutions and funding agencies that supported our research activities and international collaborations. The Erasmus+ Teaching Staff Mobility program facilitated several visits by Stefano De Marchi during which some of the core chapters of this book were conceived and developed.

Emma Perracchione acknowledges the Polytechnic of Torino, that is, her host institution, and the support of the Fondazione Compagnia di San Paolo within the framework of the Artificial Intelligence Call for Proposals, AIxtreme project (ID Rol: 71708), and the GOSSIP project ("Greedy Optimal Sampling for Solar Inverse Problems") – funded by the Ministry of University and Research (MUR) – within the PRIN 2022 program, CUP: E53C24002330001.

We are also grateful to Holly Pavelin of Cambridge University Press, who has assisted us from the beginning and provided ideas and suggestions that improved the original manuscript.

1
Introduction

Kernel-based methods represent well-established tools of modern numerical analysis and machine learning. Kernel theory has attracted ongoing interest throughout the twentieth century, with significant contributions by mathematicians such as Mercer (1909), Bochner (1933), and Aronszajn (1950), and has experienced renewed interest since the 1990s. From an approximation theory perspective, this interest was driven by the growing availability of scattered and unstructured data in scientific and engineering applications, which required meshfree approximation techniques. From another viewpoint, the development of support vector machines (SVMs) and other kernel-based learning tools demonstrated the effectiveness of kernel methods in machine learning. The strength of kernel methods lies in their flexibility and elegant theoretical grounding, which allows the construction of adaptive approximation spaces that naturally "shape" themselves to the data at hand, without requiring predetermined grids or meshes. This property makes kernel methods particularly suitable for high-dimensional problems and scattered data interpolation.

The book is organized into nineteen chapters that guide the reader from foundational concepts to recent research topics. The first part, namely Chapters 2–5, concerns fundamental notions in kernel approximation: positive-definite kernels and functions, reproducing kernel Hilbert spaces (RKHSs), polynomial precision, and error estimates. Examples of classical families of radial basis functions (RBFs) are also provided, such as Gaussians, Matérn functions, and Wendland's compactly supported functions. Chapters 6–8 explore advanced computational strategies to improve efficiency and stability: rational kernels, partition of unity (PU) methods, and efficient greedy/knot-removal algorithms. These techniques address practical challenges arising from large datasets or ill-conditioned systems.

Chapters 9–11 deal with variably scaled (discontinuous) kernels, deepening the theoretical analysis of the discontinuous setting. This novel approach enables the effective reconstruction of functions with jumps, overcoming the classical Gibbs phenomenon that plagues standard methods. Applications to magnetic particle imaging (MPI) and astronomical imaging demonstrate the practical relevance of these theoretical developments.

Then, Chapters 12–16 shift the focus to machine learning, covering SVMs and data-driven optimization techniques. In particular, we present recent research on approximation methods empowered by learning tools such as data-driven greedy approximation, discontinuous neural networks (δNNs) for learning optimal scaling functions in the variably scaled kernel framework and machine learning approaches enriched by approximation techniques such as PU-based SVMs, learning with Variably Scaled Kernels (VSKs).

Chapter 17 concerns concepts of topological data analysis and related kernels that address the output of persistent homology.

An outline of helpful information and links to available software packages related to the topics discussed throughout the book is provided in Chapter 18, where the reader can find guidance on tools for implementing kernel-based methods, performing numerical experiments, and exploring further resources.

Finally, Chapter 19 offers concluding remarks, summarizing the book's main contributions and commenting on several relevant topics that, while closely connected to kernel methods, have been intentionally omitted or only briefly mentioned because they fall outside the scope of these notes or are already thoroughly covered in other specialized references.

The book assumes familiarity with fundamental functional analysis, linear algebra, and numerical methods. Knowledge of approximation theory fundamentals is helpful, but not strictly necessary, as key concepts are introduced when needed.

Kernel methods remain a vibrant area of research with many open questions. The interplay between approximation theory, machine learning, and computational mathematics continues to generate new insights and techniques. This book provides both a solid foundation in established results and a window into current research trends, stimulating interested readers to contribute to future developments in this exciting field.

2
Positive Definite Kernels and Functions

This chapter provides a brief introduction to positive definite kernels and the relationships among their various characterizations. It discusses how radial kernels and their associated RKHSs can serve as an effective and flexible tool for addressing the scattered data interpolation problem. We omit here the proofs of well-known results, as they can be found in other fundamental books, for example, Buhmann (2003), Wendland (2005), and Fasshauer (2007).

2.1 Preliminaries and Motivations

We address the problem of reconstructing an unknown function f from a set of data. These data consist of two sets: the *data sites* or *nodes* or *points* $X = X_N = \{\boldsymbol{x}_1, \dots, \boldsymbol{x}_N\}$ arbitrarily distributed on a domain $\Omega \subseteq \mathbb{R}^d$, $d \geq 1$, and the real *data values* or *function values* $f_k = f(\boldsymbol{x}_k)$, $k = 1, \dots, N$. The reconstruction aims to approximate the (unknown) function values at locations other than the data sites. We then look for a function s that either *interpolates* the data, that is, it satisfies the conditions $s(\boldsymbol{x}_k) = f_k$, $1 \leq k \leq N$, or *approximates* the data, that is, $s(\boldsymbol{x}_k) \approx f_k$. The latter is relevant when, for instance, the samples are noisy measurements. Note that the values f_k do not necessarily originate from a function f, but we shall keep this in mind for reasons that will become clearer later. In many cases, the data are *scattered*, that is, they have no special structure, and *meshfree* approaches are then worth considering.

Let us give a more formal introduction to the interpolation problem.

Problem 2.1.1 (scattered data interpolation) *Let $\Omega \subset \mathbb{R}^d$ and let $X_N := \{\boldsymbol{x}_k, k = 1, \dots, N\} \subset \Omega$ be a set of distinct nodes, where $N, d \in \mathbb{N}$ and $N, d \geq 1$. Given an unknown function $f : \Omega \longrightarrow \mathbb{R}$, we want to approximate it from the set*

of function values $F_N := \{f_k = f(\boldsymbol{x}_k),\ k = 1,\dots,N\}$ *by imposing interpolation conditions.*
More precisely, we look for a function $s : \Omega \longrightarrow \mathbb{R}$ *such that*

$$s(\boldsymbol{x}_k) = f(\boldsymbol{x}_k)$$

for all $\boldsymbol{x}_k \in X_N$.

In order to construct the interpolant s, which we may denote as s_f, s_X, or $s_{f,X}$ when convenient, one usually considers a basis $b_1,\dots,b_N$ of functions on Ω and looks for $s \in \mathscr{B}_N := \operatorname{span}\{b_1,\dots,b_N\}$. Under this assumption, the function s can be expressed as the linear combination

$$s(\boldsymbol{x}) = \sum_{k=1}^{N} c_k b_k(\boldsymbol{x}), \quad \boldsymbol{x} \in \Omega.$$

Let A be the related *interpolation matrix*, whose entries are

$$\mathsf{A}_{ij} := b_i(\boldsymbol{x}_j), \tag{2.1}$$

and let $\boldsymbol{f} := (f_1,\dots,f_N)^{\mathsf{T}}$. As long as A is non-singular, the vector of coefficients $\boldsymbol{c} := (c_1,\dots,c_N)^{\mathsf{T}}$ is uniquely defined as the solution of the linear system $\mathsf{A}\boldsymbol{c} = \boldsymbol{f}$, thus the interpolant s is unique too.
The well-posedness of Problem 2.1.1 is guaranteed as long as $\mathscr{B}_N$ is a *Haar space.*

Definition 2.1.1 Let $\mathscr{B}_N \subseteq \mathscr{C}(\Omega)$, $N \in \mathbb{N}$, $N \geq 1$, be a finite-dimensional linear space such that $\mathscr{B}_N = \operatorname{span}\{b_1,\dots,b_N\}$. Then, $\mathscr{B}_N$ is a *Haar space* on Ω if, for any set of distinct nodes $X_N = \{\boldsymbol{x}_k,\ k = 1,\dots,N\} \subseteq \Omega$,

$$\det \mathsf{A} \neq 0,$$

where A is the interpolation matrix defined in (2.1).
The set $\{b_1,\dots,b_N\}$ is called a *Haar system.*

In the univariate case $d = 1$, an example of Haar space for a set of N distinct nodes is the space of polynomials of degree $N-1$. Unfortunately, in the multivariate case, we have Theorem 2.1.1 defined by Curtis (1959), Haar (1918), and Mairhuber (1956).

Theorem 2.1.1 (Haar–Mairhuber–Curtis) *Let* $d \geq 2$ *and let* $\Omega \subseteq \mathbb{R}^d$ *be such that it contains an interior point. Then, there exist no Haar spaces of continuous functions except for trivial ones, that is, spaces spanned by a single function.*

Proof. Let $d \geq 2$ and assume that $\mathscr{B}_N$ is a Haar space with basis $\{b_1,\ldots,b_N\}$ with $N \geq 2$. We show that this leads to a contradiction. In fact, let $\boldsymbol{x}_1,\ldots,\boldsymbol{x}_N$ be a set of N distinct points on $\Omega \subset \mathbb{R}^d$ and A the matrix such that $(\mathsf{A})_{ij} = b_i(\boldsymbol{x}_j)$, $i,j=1,\ldots,N$. By the Definition 2.1.1 of Haar space $\det(\mathsf{A}) \neq 0$. Now, consider a closed path in Ω connecting only $\boldsymbol{x}_1$ and $\boldsymbol{x}_2$. This is possible since Ω by assumption contains an interior point. We can then exchange $\boldsymbol{x}_1$ and $\boldsymbol{x}_2$ by continuously moving them along such a path (without interfering with other points $\boldsymbol{x}_i$). This means that columns 1 and 2 of the matrix A have changed, and therefore the determinant has changed sign. Since the determinant is a continuous function of $\boldsymbol{x}_1$ and $\boldsymbol{x}_2$, we must have had $\det(\mathsf{A}) = 0$ at some point along the path. This contradicts the fact that $\det(\mathsf{A}) \neq 0$. □

As a direct consequence, the Haar–Mairhuber–Curtis Theorem states that the space $\mathscr{B}_N$ cannot be unrelated to the set of data sites X_N to obtain a well-posed problem. In this direction, the notion of *unisolvent sets* is fundamental.

Definition 2.1.2 The set of points $X_N = \{\boldsymbol{x}_k,\ k=1,\ldots,N\} \subset \Omega$ is unisolvent for the function space $\mathscr{B}_N = \mathrm{span}\{b_1,\ldots,b_N\}$ if all the elements in $\mathscr{B}_N$ are completely determined by their values at X_N.

As an example, it is not possible to perform *unique* interpolation with (multivariate) polynomials of degree N to data given at arbitrary locations in $\mathbb{R}^2$, or in $\mathbb{R}^d$ for any $d \geq 2$ (see Chapter 3).

2.2 Kernels and Positive Definite Functions

We introduce kernel-based interpolation as a natural way to choose a function space that is tailored to the set of nodes. To our aims, a kernel κ on $\Omega \times \Omega \subseteq \mathbb{R}^{d\times d}$ is a symmetric real-valued function of two variables:

$$\kappa : \Omega \times \Omega \longrightarrow \mathbb{R}, \quad (\boldsymbol{x},\boldsymbol{y}) \longmapsto \kappa(\boldsymbol{x},\boldsymbol{y}).$$

Fixed $\boldsymbol{y} \in \Omega$, the univariate function $\kappa_{\boldsymbol{y}}(\boldsymbol{x}) := \kappa(\boldsymbol{x},\boldsymbol{y})$ is well-defined on Ω. Therefore, resuming the thread of the introductory Section 2.1, we can associate the space $\mathscr{X}_\kappa(\Omega) := \mathrm{span}\{\kappa(\cdot,\boldsymbol{x}_1),\ldots,\kappa(\cdot,\boldsymbol{x}_N)\}$ to the set of nodes X_N.

In this setting, the candidate interpolant of a given function f at the nodes X_N is of the form

$$s(\boldsymbol{x}) = \sum_{k=1}^{N} c_k \kappa(\boldsymbol{x},\boldsymbol{x}_k), \quad \boldsymbol{x} \in \Omega, \tag{2.2}$$

where the vector of coefficients $\boldsymbol{c}$ is possibly determined by solving the linear system

$$\mathsf{K}\boldsymbol{c} = \boldsymbol{f}, \tag{2.3}$$

with $\boldsymbol{f}$ being the vector of function values in F_N and K being the interpolation matrix (also called *kernel matrix*) defined as

$$\mathsf{K}_{ij} := \kappa(\boldsymbol{x}_i, \boldsymbol{x}_j). \tag{2.4}$$

In Sections 2.2.1–2.2.3, we analyze the hypotheses under which the kernel matrix (2.4) is invertible, which implies the well-posedness of the interpolation problem. In doing so, we consider positive-definite radial kernels. In particular, we will see that, for strictly positive definiteness, the set of nodes X_N is always unisolvent with respect to $\mathscr{X}_\kappa(\Omega)$.

After that, we investigate the properties of interpolation spaces such as $\mathscr{X}_\kappa(\Omega)$, leading to the theory of RKHSs.

2.2.1 Kernel-Based Interpolation

From now on, we adopt the abridged notation $X = X_N$ when it is possible or convenient to omit the cardinality of the set.

The uniqueness of the interpolant in (2.2) is guaranteed as long as the kernel matrix K is non-singular. A sufficient condition for the invertibility of K is the positiveness of all its eigenvalues. This observation leads to Definition 2.2.1.

Definition 2.2.1 A real symmetric matrix $\mathsf{A} \in M^{N \times N}$ is called positive definite if the associated quadratic form is positive, that is,

$$\boldsymbol{c}^\intercal \mathsf{A} \boldsymbol{c} > 0, \tag{2.5}$$

for any nonzero $\boldsymbol{c} = (c_1, \ldots, c_N)^\intercal \in \mathbb{R}^N$. Moreover, A is positive semi-definite if the quadratic form (2.5) is nonnegative.

Therefore, we are interested in kernels whose associated kernel matrices are positive (semi-)definite for any choice of node set X. These kernels are the so-called *(strictly) positive definite* kernels.

Definition 2.2.2 A kernel κ is strictly positive definite on $\Omega \times \Omega \subset \mathbb{R}^{d \times d}$ if its associated kernel matrix K, defined in (2.4), is positive definite for any set of nodes $X \subset \Omega$. If K is positive semi-definite, then κ is positive definite.

We point out that in Fasshauer (2007), positive definite kernels are associated with positive semi-definite matrices, while positive definite matrices are linked to strictly positive definite kernels. However, more recently in Fasshauer and McCourt (2015), this nomenclature is often relaxed by matching positive definite kernels and positive definite matrices. Here, we adopt the terminology of Fasshauer (2007). Furthermore, even if the invertibility of the kernel matrix is guaranteed only in the strictly positive definite case, we will see that kernel-based interpolation can be performed using kernels that are only positive definite.

Along with Definition 2.2.2, we introduce the notion of (strictly) *positive definite functions* associated with *translation invariant* kernels.

Definition 2.2.3 A kernel $\kappa : \Omega \times \Omega \longrightarrow \mathbb{R}$ is translation invariant if there exists a function $\phi : \Omega \longrightarrow \mathbb{R}$ such that

$$\kappa(\boldsymbol{x},\boldsymbol{y}) = \phi(\boldsymbol{x}-\boldsymbol{y}), \quad \boldsymbol{x},\boldsymbol{y} \in \Omega.$$

Moreover, we say that ϕ is a (strictly) positive definite function if κ is (strictly) positive definite.

Translation invariant kernels are also referred to as *convolution kernels*, and they play a central role in the theory of integral equations.

Although we will restrict ourselves to real-valued kernels, we now consider the complex-valued setting to present some properties of positive definite functions and thus of translation invariant kernels (Fasshauer, 2007, section 3.1). Extending our analysis to complex-valued positive definite kernels and functions is straightforward. Indeed, with respect to the quadratic form in (2.5), we only need to take $\boldsymbol{c} \in \mathbb{C}^N$ and then consider the Hermitian quadratic form $\boldsymbol{c}^{\mathsf{H}} \mathsf{A} \boldsymbol{c}$, where $\boldsymbol{c}^{\mathsf{H}} = \overline{\boldsymbol{c}}^{\mathsf{T}}$ denotes the conjugate transpose.

Theorem 2.2.1 *Let $\phi : \mathbb{R}^d \longrightarrow \mathbb{C}$ be a positive definite function. Then:*

(i) $\phi(\mathbf{0}) \geq 0$.
(ii) $\phi(-\boldsymbol{x}) = \overline{\phi(\boldsymbol{x})}$.
(iii) Since $|\phi(\boldsymbol{x})| \leq \phi(\mathbf{0})$, ϕ *is bounded.*
(iv) If $\phi(\mathbf{0}) = 0$, *then* $\phi \equiv 0$.
(v) Nonnegative finite linear combinations of positive definite functions are positive definite, that is, if $\phi_1, \ldots, \phi_N$ *are positive definite on* $\mathbb{R}^d$ *and* $c_k \geq 0$, $k = 1, \ldots, N$, *then*

$$\phi(\boldsymbol{x}) = \sum_{k=1}^{N} c_k \phi_k(\boldsymbol{x}), \quad \boldsymbol{x} \in \mathbb{R}^d,$$

is also positive definite. Moreover, if at least one of the ϕ_k is strictly positive definite and the corresponding $c_k > 0$, then ϕ is strictly positive definite.

(vi) The product of (strictly) positive definite functions is (strictly) positive definite.

Remark 2.2.1 *The properties stated in Theorem 2.2.1 suggest that the positive definite function may be bell-shaped* (see Figure 2.1 for a popular univariate example: the Gaussian function, here plotted in the interval $[-1,1]$)..

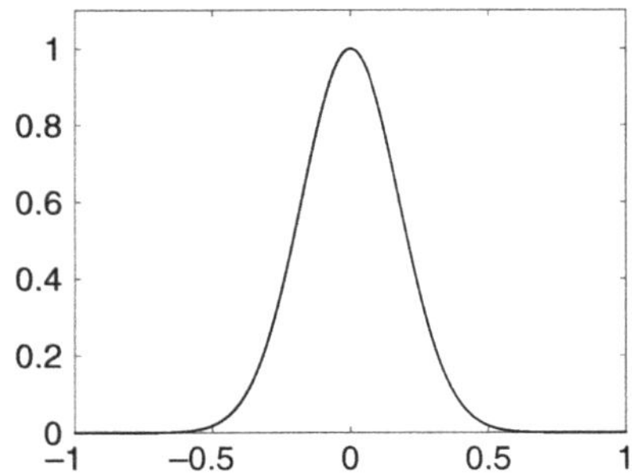

Figure 2.1 The bell-shaped function $f(x) = \mathrm{e}^{-4x^2}$.

We fix some notations. We denote the Euclidean scalar product in $\mathbb{R}^d$ by

$$\boldsymbol{x} \cdot \boldsymbol{y} := \boldsymbol{x}^\mathsf{T} \boldsymbol{y}, \quad \boldsymbol{x}, \boldsymbol{y} \in \mathbb{R}^d.$$

Moreover, given $\Omega \subseteq \mathbb{R}^d$ and $p \geq 1$, we define the well-known function space

$$\mathscr{L}_p := \left\{ f : \Omega \longrightarrow \mathbb{R} \,\middle|\, \int_\Omega |f(\boldsymbol{x})|^p \mathrm{d}\boldsymbol{x} < +\infty \right\},$$

equipped with the norm

$$\|f\|_p := \left(\int_\Omega |f(\boldsymbol{x})|^p \mathrm{d}\boldsymbol{x} \right)^{1/p}.$$

Remark 2.2.2 *For every $p \geq 1$, $\mathscr{L}_p$ is a Banach space, that is, it is normed and complete. Moreover, as a relevant case, if $p = 2$ then $\mathscr{L}_2$ is a Hilbert space endowed with the scalar product*

$$\langle f, g \rangle_{\mathscr{L}_2} := \int_\Omega f(\boldsymbol{x}) g(\boldsymbol{x}) \mathrm{d}\boldsymbol{x},$$

where $f, g \in \mathscr{L}_2$.

Here, we present some integral characterizations of complex-valued positive definite functions. To this end, we first recall Definition 2.2.4.

Definition 2.2.4 Let $f \in \mathscr{L}_1(\mathbb{R}^d)$ and let μ be a finite measure on $\mathbb{R}^d$.

(i) The d-dimensional Fourier transform of f is

$$(\mathrm{F}f)(\boldsymbol{\omega}) := \frac{1}{\sqrt{(2\pi)^d}} \int_{\mathbb{R}^d} f(\boldsymbol{x})\, e^{-i\boldsymbol{\omega}\cdot\boldsymbol{x}} \mathrm{d}\boldsymbol{x}, \quad \boldsymbol{\omega} \in \mathbb{R}^d.$$

(ii) The d-dimensional Fourier transform of μ is

$$(\mathrm{F}\mu)(\boldsymbol{\omega}) := \frac{1}{\sqrt{(2\pi)^d}} \int_{\mathbb{R}^d} e^{-i\boldsymbol{\omega}\cdot\boldsymbol{x}} \mathrm{d}\mu(\boldsymbol{x}), \quad \boldsymbol{\omega} \in \mathbb{R}^d.$$

Positive definite functions are fully characterized in terms of Fourier transforms. This important result is stated in the well-known Bochner Theorem, whose original proof can be found in Bochner (1933).

Theorem 2.2.2 (Bochner) *Let $\phi : \mathbb{R}^d \longrightarrow \mathbb{C}$ be a continuous function. Then, ϕ is positive definite on $\mathbb{R}^d$ if and only if it is the d-dimensional Fourier transform of a finite nonnegative Borel measure μ on $\mathbb{R}^d$.*

By adding a further hypothesis, Bochner Theorem is restricted to strictly positive definite functions. To show this, we need to introduce Definition 2.2.5.

Definition 2.2.5 Let μ be a nonnegative Borel measure defined on some topological space X. The *carrier* of μ is given by

$$X \setminus \cup \{O : O \text{ is open and } \mu(O) = 0\}.$$

Then, we can state the sufficient condition as in Theorem 2.2.3 for strictly positive definite functions (Fasshauer, 2007, section 3.2).

Theorem 2.2.3 *Let μ be a nonnegative finite Borel measure on $\mathbb{R}^d$, whose carrier is a set of nonzero Lebesgue measure. Then, the d-dimensional Fourier transform of μ is strictly positive definite on $\mathbb{R}^d$.*

Corollary 2.2.1 *Let $\phi \in \mathscr{L}_1(\mathbb{R}^d)$ be a continuous nonnegative function that is not identically zero. Then, the d-dimensional Fourier transform of ϕ is strictly positive definite on $\mathbb{R}^d$.*

Finally, if ϕ is bounded, we can recover a sufficient and necessary condition for its strictly positive definiteness.

Theorem 2.2.4 *Let $\phi \in \mathscr{L}_1(\mathbb{R}^d)$ be a continuous function. Then, ϕ is strictly positive definite if and only if ϕ is bounded and its d-dimensional Fourier transform is nonnegative and not identically equal to zero.*

2.2.2 Positive Definite Radial Functions

Generally, the complexity of Problem 2.1.1 depends on the dimension in which the data lie and the underlying function is defined. For example, multivariate polynomial interpolation presents many more difficulties than the univariate case. In view of this, it is preferable to consider kernel functions that are somehow blind to the dimension of the underlying space. Therefore, we restrict our investigation to positive definite kernels that are *radial* in the Euclidean sense, as explained in Definition 2.2.6.

Definition 2.2.6 A kernel $\kappa : \Omega \times \Omega \longrightarrow \mathbb{R}$ is radial if there exists a univariate function $\varphi : [0,\infty) \longrightarrow \mathbb{R}$ such that

$$\kappa(\boldsymbol{x},\boldsymbol{y}) = \varphi(||\boldsymbol{x}-\boldsymbol{y}||) = \varphi(r),$$

where $r = ||\boldsymbol{x}-\boldsymbol{y}||$ and $||\cdot|| = ||\cdot||_2$ is the Euclidean norm.
In this case, we refer to φ as the univariate function associated with ϕ and κ.

We point out that other norms could be taken into account in Definition 2.2.6. However, the standard Euclidean norm is suitable for our purposes.

Remark 2.2.3 *Radial kernels are translation invariant. Thus, we refer to functions ϕ and φ as (strictly) positive definite radial functions, inheriting the properties of κ. In particular, φ is called a RBF, and, in this sense, the kernel-based interpolation setting we are presenting is often referred to as RBF interpolation.*

Lemma 2.2.1 extends the dimensional blindness of radial functions to the positive definite setting.

Lemma 2.2.1 *Let ϕ be (strictly) positive definite and radial on $\mathbb{R}^d$. Then, ϕ is also (strictly) positive definite and radial on $\mathbb{R}^{\tilde{d}}$ with $\tilde{d} \leq d$.*

Similarly to what was discussed in Section 2.2.1, we present an integral characterization of (strictly) positive definite radial functions referring to Fasshauer (2007, section 3.3).

Definition 2.2.7 Let $\varphi : [0,\infty) \longrightarrow \mathbb{R}$ be a continuous radial function. The Hankel transform of order d of φ is defined as

$$(\mathrm{H}\varphi)(r) := \frac{1}{\sqrt{r^{d-2}}} \int_0^\infty \varphi(t)\, t^{d/2} J^1_{(d-2)/2}(rt)\, \mathrm{d}t,$$

where $J^1_{(d-2)/2}$ is the classical Bessel function of the first kind of order $(d-2)/2$.

Definition 2.2.8 Let μ be a finite nonnegative Borel measure on $[0,\infty)$. The Bessel transform of order d of μ is defined as

$$(\mathrm{B}\mu)(r) := \int_0^\infty D_d(rt)\,\mathrm{d}\mu(t),$$

where

$$D_d(r) := \begin{cases} \cos r, & \text{for } d=1, \\ \Gamma\left(\frac{d}{2}\right)\left(\frac{2}{r}\right)^{(d-2)/2} J^1_{(d-2)/2}(r), & \text{for } d \geq 2 \end{cases}$$

where $J^1_{(d-2)/2}$ is the classical Bessel function of the first kind of order $(d-2)/2$.

Theorem 2.2.5 can be found in Wendland (2005, section 5.2).

Theorem 2.2.5 *Let $\phi \in \mathscr{L}_1(\mathbb{R}^d)$ be a continuous and radial function, with associated univariate function φ. Then, its d-dimensional Fourier transform* $\mathrm{F}\phi$ *is also radial. Moreover, the univariate function associated with* $\mathrm{F}\phi$ *is* $\mathrm{H}\varphi$*, which is the Hankel transform of order d of φ.*

We can now state the integral characterizations of strictly positive definite functions as given in Theorem 2.2.6. The first was originally formulated in Schoenberg (1938).

Theorem 2.2.6 *A continuous function $\varphi : [0,\infty) \longrightarrow \mathbb{R}$ is positive definite and radial on $\mathbb{R}^d$ if and only if it is the Bessel transform of order d of a finite nonnegative Borel measure μ on $[0,\infty)$.*

Theorem 2.2.7 *A continuous radial function $\varphi : [0,\infty) \longrightarrow \mathbb{R}$, such that $r \to r^{d-1}\phi(r) \in \mathscr{L}_1[0,\infty)$, is strictly positive definite and radial on $\mathbb{R}^d$ if and only if its Hankel transform of order d is nonnegative and not identically equal to zero.*

The characterization of functions that are strictly positive definite on $\mathbb{R}^d$ for all d is provided by Theorem 2.2.8 in Schoenberg (1938) and Micchelli (1986).

Theorem 2.2.8 (Schoenberg) *Let $\varphi : [0,\infty) \longrightarrow \mathbb{R}$ be a continuous function. Then, φ is strictly positive definite and radial on $\mathbb{R}^d$ for all d if and only if*

$$\varphi(r) = \int_0^\infty e^{-r^2t^2}\,\mathrm{d}\mu(t), \tag{2.6}$$

where μ is a finite nonnegative Borel measure on $[0,\infty)$ not concentrated at the origin.

Theorem 2.2.8 has interesting direct consequences. Basically, being μ a non-negative measure and being the exponential in (2.6) positive, the function φ never vanishes unless μ is the zero measure. Thus, we have Corollary 2.2.2.

Corollary 2.2.2 *There are no continuous univariate oscillatory functions that are strictly positive definite and radial on $\mathbb{R}^d$ for all d. Moreover, there are no compactly supported univariate continuous functions that are strictly positive definite and radial on $\mathbb{R}^d$ for all d.*

Although RBFs are suitable for being taken into account in any dimension, specific techniques for high-dimensional settings and manifolds have been investigated in literature (Albrecht et al., 2025; Avesani et al., 2025; Hangelbroek and Rieger, 2025; Kempf and Wendland, 2023; and Rieger and Wendland, 2024).

2.2.3 Completely and Multiply Monotone Functions

Strictly positive definite radial functions can also be characterized via the so-called *completely* and *multiply monotone functions*. This turns out to be useful, especially when the Fourier transform is not easy to compute. For the following characterizations and results, we refer to Fasshauer (2007, chapter 5).

Definition 2.2.9 A function $\varphi : [0,\infty) \longrightarrow \mathbb{R}$, $\varphi \in \mathscr{C}[0,\infty) \cap \mathscr{C}^{\infty}(0,\infty)$ that satisfies

$$(-1)^k \varphi^{(k)}(r) \geq 0, \; r > 0, \; k = 0, 1, 2, \dots$$

is called completely monotone (CM) on $[0,\infty)$.

This definition allows to verify when a function φ is (strictly) positive definite and radial for all dimensions d.

Theorem 2.2.9 *A function φ is* CM *on $[0,\infty)$ if and only if $\bar{\varphi} = \varphi(\|\cdot\|^2)$ is positive definite and radial on $\mathbb{R}^d$ for all d.*

Theorem 2.2.10 *A function φ is* CM *on $[0,\infty)$ but not constant if and only if $\bar{\varphi} = \varphi(\|\cdot\|^2)$ is strictly positive definite and radial on $\mathbb{R}^d$ for all d.*

A further characterization in terms of *multiply monotone* functions allows us to understand when a function is strictly positive definite and radial on $\mathbb{R}^d$ for some fixed d.

Definition 2.2.10 A function $\varphi : (0,\infty) \longrightarrow \mathbb{R}$, $\varphi \in \mathscr{C}^{k-2}(0,\infty)$, $k \geq 0$ and for which $(-1)^t \varphi^{(t)}(r)$ is nonnegative, nonincreasing, and convex for $t = 0, 1, \ldots, k-2$ is called k-times monotone on $(0,\infty)$. In the case $k = 1$, we only require $\varphi \in \mathscr{C}(0,\infty)$ to be nonnegative and nonincreasing.

Convexity can be expressed by the inequality $\varphi'' \geq 0$, if φ'' exists. Hence, a multiply monotone function is essentially a CM function with *truncated monotonicity*. Then Theorem 2.2.11 holds.

Theorem 2.2.11 (Micchelli) *Let $k = \lfloor t/2 \rfloor + 2$ be a positive integer. If $\varphi : [0,\infty) \longrightarrow \mathbb{R}$ is continuous and k-times monotone on $[0,\infty)$, but not constant, then, φ is strictly positive definite and radial on $\mathbb{R}^t$ for any t such that $\lfloor t/2 \rfloor \leq k-2$.*

2.3 Reproducing Kernel Hilbert Spaces

In the introduction of Section 2.2, we presented the need to study the well-posedness of Problem 2.1.1 along with the associated interpolation space $\mathscr{X}_\kappa$. While we already investigated the former and consequently considered (strictly) positive-definite radial functions and kernels, in Section 2.3, we discuss the latter.

We begin our analysis by presenting the definition of RKHS, first introduced in Aronszajn (1950).

Definition 2.3.1 Let $\mathscr{H}(\Omega) = \mathscr{H}$ be a Hilbert space of real-valued functions defined on $\Omega \subset \mathbb{R}^d$. Then, $\mathscr{H}$ is an RKHS if there exists a function $\kappa : \Omega \times \Omega \longrightarrow \mathbb{R}$, also called the *reproducing kernel*, such that

(i) Let $\boldsymbol{x} \in \Omega$ and $\kappa_{\boldsymbol{x}}(\boldsymbol{y}) := \kappa(\boldsymbol{x}, \boldsymbol{y})$, with $\boldsymbol{y} \in \Omega$. Then $\kappa_{\boldsymbol{x}} \in \mathscr{H}$.
(ii) Let $\langle \cdot, \cdot \rangle_{\mathscr{H}}$ be the scalar product associated with $\mathscr{H}$. We have

$$f(\boldsymbol{x}) = \langle f, \kappa_{\boldsymbol{x}} \rangle_{\mathscr{H}}$$

for any $f \in \mathscr{H}$ and $\boldsymbol{x} \in \Omega$. This is called the reproducing property.

We also denote by $\|\cdot\|_{\mathscr{H}} = \sqrt{\langle \cdot, \cdot \rangle_{\mathscr{H}}}$ the Hilbert norm.

Two important results concerning the uniqueness and existence of the reproducing kernel are stated (Fasshauer and McCourt, 2015, section 2.3).

Proposition 2.3.1 *The reproducing kernel κ in an RKHS $\mathscr{H}$ is unique.*

Proposition 2.3.2 *Let $\mathscr{H}$ be a Hilbert space of real-valued functions defined on $\Omega \subset \mathbb{R}^d$ and let $\delta_{\boldsymbol{x}} : \mathscr{H} \longrightarrow \mathbb{R}$ be the point evaluation associated with $\boldsymbol{x} \in \Omega$. Then, $\delta_{\boldsymbol{x}}$ is continuous on Ω if and only if $\mathscr{H}$ is a RKHS.*

The uniqueness of the reproducing kernel allows us to write $\mathscr{H} = \mathscr{H}_\kappa$ if $\mathscr{H}$ is an RKHS, also denoting the reproducing kernel. The result in Proposition 2.3.2 is based on the well-known *Riesz representation Theorem* (Riesz, 1909).

The reproducing property further implies other results reported in Proposition 2.3.3 (see Guillemard and Iske, 2017, and Fasshauer, 2007, section 13.1).

Proposition 2.3.3 *Let $\mathscr{H}_\kappa$ be an RKHS on $\Omega \subset \mathbb{R}^d$.*

(i) *The kernel κ is symmetric.*
(ii) $\kappa(\boldsymbol{x},\boldsymbol{y}) = (\kappa(\cdot,\boldsymbol{y}), \kappa(\cdot,\boldsymbol{x}))_{\mathscr{H}_\kappa}$, *for $\boldsymbol{x},\boldsymbol{y} \in \Omega$.*
(iii) *The convergence in the Hilbert space norm implies pointwise convergence, i.e., if $\|f - f_n\|_{\mathscr{H}_\kappa} \longrightarrow 0$ for $n \longrightarrow \infty$, then $|f(\boldsymbol{x}) - f_n(\boldsymbol{x})| \longrightarrow 0$ for all $\boldsymbol{x} \in \Omega$.*

A consequence of Propositions 2.3.2 and 2.3.3 is that any linear subspace of an RKHS is itself an RKHS (see Guillemard and Iske, 2017).

Theorem 2.3.1 connects the reproducing kernel of an RKHS to the previously investigated positive definite kernels (see Definition 2.2.2) (Fasshauer, 2007, section 13.1).

Theorem 2.3.1 *Let $\mathscr{H}_\kappa$ be an RKHS on $\Omega \subset \mathbb{R}^d$ and let $\mathscr{H}_\kappa^*$ be the dual space of bounded linear functionals on $\mathscr{H}_\kappa$. Then, κ is positive definite on $\Omega \times \Omega$. Moreover, κ is strictly positive definite if and only if for every $\boldsymbol{x} \in \Omega$ the point evaluation functional $\delta_{\boldsymbol{x}}$ is linearly independent in $\mathscr{H}_\kappa^*$.*

In summary, we introduced the RKHS and linked the associated reproducing kernel to strictly positive definite kernels, which have been discussed in the context of the uniqueness of the solution to Problem 2.1.1. Actually, we are more interested in the other direction. That is, we fix a strictly positive definite kernel κ and want to characterize the interpolation space $\mathscr{X}_\kappa$ as an RKHS.

Let $\Omega \subset \mathbb{R}^d$ and let κ be a strictly positive definite kernel on $\Omega \times \Omega$. We define the following space,

$$\widetilde{\mathscr{N}_\kappa}(\Omega) = \mathrm{span}\{\kappa(\cdot,\boldsymbol{x}), \, \boldsymbol{x} \in \Omega\},$$

which can be equipped with the bilinear form $(\cdot,\cdot)_{\widetilde{\mathscr{N}_\kappa}}(\Omega)$ defined as

$$\left\langle \sum_{i=1}^{N} c_i \kappa\langle\cdot,\boldsymbol{x}_i\rangle, \sum_{j=1}^{N} d_j \kappa(\cdot,\boldsymbol{x}_j) \right\rangle_{\widetilde{\mathscr{N}_\kappa}(\Omega)} = \sum_{i=1}^{N}\sum_{j=1}^{N} c_i d_j \kappa(\boldsymbol{x}_i,\boldsymbol{x}_j).$$

Theorem 2.3.2 characterizes the so-constructed function space (Fasshauer, 2007, section 13.2), which in turn allows us to introduce the notion of *native space* (Schaback, 2000).

Theorem 2.3.2 *Let κ be a strictly positive definite kernel on $\Omega \times \Omega$. Then, the bilinear form $\langle \cdot, \cdot \rangle_{\widetilde{\mathcal{N}_\kappa}(\Omega)}$ defines an inner product on $\widetilde{\mathcal{N}_\kappa}(\Omega)$. Moreover, $\widetilde{\mathcal{N}_\kappa}(\Omega)$ is a pre-Hilbert space with reproducing kernel κ.*

Definition 2.3.2 Let κ be a strictly positive definite kernel on $\Omega \times \Omega$. The native space $\mathcal{N}_\kappa(\Omega)$ of κ is the completion of $\widetilde{\mathcal{N}_\kappa}(\Omega)$ with respect to the norm $\|\cdot\|_{\widetilde{\mathcal{N}_\kappa}(\Omega)}$, so that $\|f\|_{\widetilde{\mathcal{N}_\kappa}(\Omega)} = \|f\|_{\mathcal{N}_\kappa(\Omega)}$ for all $f \in \widetilde{\mathcal{N}_\kappa}(\Omega)$.

Finally, the interpolation space $\mathscr{X}_\kappa(\Omega)$ is characterized by Proposition 2.3.4.

Proposition 2.3.4 *Let $\Omega \subset \mathbb{R}^d$ and let κ be a strictly positive definite kernel on $\Omega \times \Omega$, with associated native space $\mathcal{N}_\kappa(\Omega)$. If $X = \{\boldsymbol{x}_k,\ k = 1, \ldots, N\} \subset \Omega$ is a set of distinct nodes in Ω, then the finite linear subspace of $\mathcal{N}_\kappa(\Omega)$ defined as*

$$\mathscr{X}_\kappa(\Omega) = span\{\kappa(\cdot, \boldsymbol{x}_1), \ldots, \kappa(\cdot, \boldsymbol{x}_N)\}$$

is an RKHS with reproducing kernel κ.
Moreover, if $f : \Omega \longrightarrow \mathbb{R}$, then Problem 2.1.1 has a unique solution in $\mathscr{X}_\kappa(\Omega)$.

Native spaces are fundamental pillars to introducing error bounds for the interpolation process, as shown in Chapter 4.

3

RBF Interpolation and Polynomial Precision

The ability of an interpolant to reproduce polynomials up to a fixed degree, known as *polynomial precision*, is often a desirable property, as polynomials are easy to manipulate and possess many useful properties. This chapter focuses on the description of conditionally positive definite kernels, which exhibit the polynomial precision property.

3.1 On Multivariate Polynomial Interpolation

As we already pointed out in Chapter 2, multivariate polynomials are not suitable for solving the scattered data interpolation problem for any X_N. Indeed, letting $\mathscr{P}_m(\mathbb{R}^d)$ be the space of d-variate polynomials of total degree less than or equal to m, there are two major issues.

1. Only data sites in special locations can guarantee the well-posedness of the interpolation problem, that is, the invertibility of the interpolation matrix.
2. Any basis $P_M = \{p_1, \ldots, p_M\}$ for the space $\mathscr{P}_m(\mathbb{R}^d)$ consists of $M = \binom{m+d}{m}$ basis functions. As a consequence, the invertibility of the interpolation matrix P, where

$$\mathsf{P}_{ij} = p_j(\boldsymbol{x}_i), \quad i = 1, \ldots, N, \; j = 1, \ldots, M,$$

is possible only if $N = M$, which is very restrictive (e.g., there is no unique polynomial interpolant at $N = 14$ data sites in $\mathbb{R}^2$).

In principle, the interpolation problem can be restricted to a suitable subset of X_N of cardinality $M < N$, but this leads to a loss of information. In this chapter, we will instead require X_N to be m-unisolvent, that is, we require that the only polynomial of total degree at most m interpolating the zero data on X_N is the zero polynomial. Consequently, there exists a unique polynomial in $\mathscr{P}_m(\mathbb{R}^d)$

that interpolates a subset of cardinality M of the data X_N (c.f. Definition 2.1.2). For example, in $\mathbb{R}^2$, three collinear points are not 1-unisolvent, as a linear interpolant, which is a plane through three arbitrary heights at these three collinear points, is not uniquely determined. On the square $[-1,1]^2 \subset \mathbb{R}^2$, the Padua points form the first complete example of unisolvent points whose *Lebesgue constant* has optimal growth (Bos et al., 2006). An example of technique for constructing unisolvent sets for multivariate polynomial interpolation has been proposed in Chung and Yao (1977). Another notable unisolvent set on the square $[-1,1]^2$ for the bivariate polynomials of degree up to n, with optimal growth of the Lebesgue constant, that is, $\log^2(n)$, is the set of *Padua points* (Caliari et al., 2005; Bos et al., 2006).

3.2 Polynomial Precision

Despite the problems outlined earlier, the capability of an interpolant to reproduce polynomials up to a fixed degree, that is, the so-called *polynomial precision*, is often a desirable property. A kernel-based interpolant built upon a strictly positive definite RBF φ could indeed perform well in reconstructing *elaborated* underlying functions, but yields to poor representations of simple constant and linear functions, which are zero and one degree polynomials.

A solution consists of modifying the sought interpolant in (2.2) to include polynomial terms, that is

$$s(\boldsymbol{x}) = \sum_{k=1}^{N} c_k \kappa(\boldsymbol{x}, \boldsymbol{x}_k) + \sum_{k=1}^{M} d_k p_k(\boldsymbol{x}), \quad \boldsymbol{x} \in \Omega, \tag{3.1}$$

where $p_1, \ldots, p_M$ is a basis for the d-variate polynomials of degree less than or equal to $m-1$.

Imposing the N interpolation conditions is not sufficient for (3.1), as we enlarged the number of parameters to $N+M$. To solve this small issue, it is a common practice to impose the M side-conditions

$$\sum_{i=1}^{N} c_i p_j(\boldsymbol{x}_i) = 0, \quad j = 1, \ldots, M.$$

Example 3.2.1 *Let us assume that we can reproduce linear functions in $\mathbb{R}^2$. Then, we consider the interpolant*

$$s(\boldsymbol{x}) = \sum_{k=1}^{N} c_k \kappa(\boldsymbol{x}, \boldsymbol{x}_k) + d_1 x^1 + d_2 x^2 + d_3, \quad \boldsymbol{x} = (x^1, x^2) \in \mathbb{R},$$

with the three conditions

$$\sum_{i=1}^{N} c_i = 0, \sum_{i=1}^{N} c_i x_i^1 = 0, \sum_{i=1}^{N} c_i x_i^2 = 0.$$

To determine the coefficients in (3.1) we then need to solve the linear system of the form

$$\begin{pmatrix} \mathsf{K} & \mathsf{P} \\ \mathsf{P}^{\mathsf{T}} & \mathsf{O} \end{pmatrix} \begin{pmatrix} \boldsymbol{c} \\ \boldsymbol{d} \end{pmatrix} = \begin{pmatrix} \boldsymbol{f} \\ \boldsymbol{0} \end{pmatrix} \tag{3.2}$$

where K is the $N \times N$ kernel matrix, P is $N \times M$ polynomial interpolation matrix, O is the $M \times M$ zero matrix and $\boldsymbol{0}$ is the M-dimensional zero vector.

In Section 3.3, we present hypotheses under which the matrix in (3.2) is invertible, which implies the existence and uniqueness of the proposed interpolant with polynomial precision. As we will see, weaker assumptions on κ than strictly positive definiteness are required.

3.3 Conditionally Positive Definite Matrices and Kernels

To prove that the augmented system (3.2) is non-singular, we introduce Theorem 3.3.1.

Theorem 3.3.1 *A kernel $\kappa : \Omega \times \Omega \subseteq \mathbb{R}^{d \times d}$ is said to be conditionally positive definite of order m on $\mathbb{R}^d$ if*

$$\sum_{i=1}^{N} \sum_{j=1}^{N} c_i c_j \kappa(\boldsymbol{x}_i, \boldsymbol{x}_j) \geq 0,$$

for any set $X_N = \{\boldsymbol{x}_1, \dots, \boldsymbol{x}_N\} \subset \mathbb{R}^d$ of pairwise distinct points, and for any $\boldsymbol{c} = (c_1, \dots, c_N)^{\mathsf{T}} \subset \mathbb{R}^N \setminus \{0\}$ that satisfies

$$\sum_{k=1}^{N} c_k p(\boldsymbol{x}_k) = 0$$

for any real-valued polynomial p of degree less than or equal to $m - 1$. If $\boldsymbol{c} \neq \boldsymbol{0}$ implies strictly inequality, then κ is strictly conditionally positive definite order m. Furthermore, the kernel matrix K is said to be (strictly) conditionally positive definite of order m if so is κ.

The first important fact concerning (strictly) conditionally positive definite functions is their order. To this aim, we state the following relevant result.

Proposition 3.3.1 *A function that is (strictly) conditionally positive definite of order m is also (strictly) conditionally positive definite of any order $\tilde{m} \geq m$. Moreover, a function that is (strictly) conditionally positive definite of order m on $\mathbb{R}^d$ is also (strictly) conditionally positive definite of order m on $\mathbb{R}^k$ with $k \leq d$.*

When speaking about the order of this class of functions, we will always refer to the smallest possible m.

The special case $m = 1$ appears already in the linear algebra literature but is presented in terms of *almost negative definiteness*. The constraints are simply $\sum_{i=1}^N c_i = 0$, and the polynomial reproduction is limited to constants. Moreover, a conditionally positive definite matrix of order one is positive definite on a subspace of dimension $N-1$; therefore, it possesses at least $N-1$ positive eigenvalues. To further deepen this case, we recall Theorem 3.3.2 (Horn and Johnson, 2013, Theorem 4.2.6).

Theorem 3.3.2 (Courant–Fisher) *Let A be a real symmetric $N \times N$ matrix with eigenvalues $\lambda_1 \geq \lambda_2 \geq \cdots \geq \lambda_N$, then*

$$\lambda_k = \max_{\dim \mathscr{V}=k} \min_{\substack{\boldsymbol{x}\in\mathscr{V} \\ \|\boldsymbol{x}\|=1}} \boldsymbol{x}^\mathsf{T}\mathsf{A}\boldsymbol{x} \quad \textit{and} \quad \lambda_k = \min_{\dim \mathscr{V}=N-k+1} \max_{\substack{\boldsymbol{x}\in\mathscr{V} \\ \|\boldsymbol{x}\|=1}} \boldsymbol{x}^\mathsf{T}\mathsf{A}\boldsymbol{x},$$

where $\mathscr{V}$ denotes the set of all linear subspaces of $\mathbb{R}^N$.

With a further assumption on the matrix, which is realized by some interesting kernels, the following stronger result holds true (Fasshauer, 2007, Theorem 9.6).

Theorem 3.3.3 *An $N \times N$ symmetric matrix K that is strictly conditionally positive definite of order one and has a nonpositive trace possesses one negative and $N-1$ positive eigenvalues.*

Proof. Let $\lambda_1 \geq \lambda_2 \geq \cdots \geq \lambda_N$ denote the eigenvalues of K. Theorem 3.3.2 implies

$$\lambda_2 = \min_{\dim \mathscr{V}=N-1} \max_{\substack{\boldsymbol{x}\in\mathscr{V} \\ \|\boldsymbol{x}\|=1}} \boldsymbol{x}^\mathsf{T}\mathsf{K}\boldsymbol{x} \leq \max_{\substack{\boldsymbol{c}:\, \sum c_k=0 \\ \|\boldsymbol{c}\|=1}} \boldsymbol{c}^\mathsf{T}\mathsf{K}\boldsymbol{c} > 0.$$

Consequently, K has at least $N-1$ positive eigenvalues. However, since $\mathrm{tr}(\mathsf{K}) = \sum_{k=1}^N \lambda_k \leq 0$, K must also have at least one positive eigenvalue. □

Returning to the general setting, one can prove the following significant result (Wendland, 2005, Theorem 8.21).

Theorem 3.3.4 *Let κ be a strictly conditionally positive definite kernel of order m and assume that the data sites $\boldsymbol{x}_1, \ldots, \boldsymbol{x}_N$ form an $(m-1)$-unisolvent set. Then the system in* (3.2) *is uniquely solvable.*

Proof. Assume that the solution $(\boldsymbol{c}, \boldsymbol{d})^\mathsf{T} \in \mathbb{R}^{N+M}$ is a solution of the homogeneous system, that is, with $\boldsymbol{f} = \boldsymbol{0}$. We then prove that $\boldsymbol{c} = \boldsymbol{0}$ and $\boldsymbol{d} = \boldsymbol{0}$. From the first block, multiplying by $\boldsymbol{c}^\mathsf{T}$, we have

$$\boldsymbol{c}^\mathsf{T} \mathsf{K} \boldsymbol{c} + \boldsymbol{c}^\mathsf{T} \mathsf{P} \boldsymbol{d} = 0,$$

and using the second equation $\mathsf{P}^\mathsf{T} \boldsymbol{c} = \boldsymbol{c}^\mathsf{T} \mathsf{P} = \boldsymbol{0}^\mathsf{T}$ we get $\boldsymbol{c}^\mathsf{T} \mathsf{K} \boldsymbol{c} = 0$. Since K is conditionally positive definite of order m, by assumption we get $\boldsymbol{c} = \boldsymbol{0}$. The unisolvency of the data sites, that is, P has linearly independent columns, and the fact that $\boldsymbol{c} = \boldsymbol{0}$ guarantees $\boldsymbol{d} = \boldsymbol{0}$ from the top block $\mathsf{K}\boldsymbol{c} + \mathsf{P}\boldsymbol{d} = \boldsymbol{0}$. □

Conditionally positive definite radial kernels can be characterized by means of generalized Fourier transforms, as well as via completely and multiply monotone functions (see Chapter 4 for examples of completely and multiply monotone functions). In this book, we do not treat these aspects that have been presented in, for example, Fasshauer (2007, chapter 9) and Wendland (2005, chapter 8).

4

Error Estimates and Examples of Kernels

This chapter provides an overview of classical error estimates for kernel-based interpolation. We remark that more elaborate error estimates have been studied in the literature, leveraging, for example, the so-called *sampling inequalities* (see, e.g., Narcowich et al., 2005; Rieger and Zwicknagl, 2010; and Wendland and Rieger, 2005). This approach will be helpful in Chapter 9. However, our purpose here is to present the basic tools that underpin RBF approximation theory. We will first focus on the theoretical aspects and then move toward empirical strategies to estimate the error in practical applications. Here, we restrict our analysis to strictly positive definite and radial kernels, as they guarantee a well-posed problem without additional constraints.

4.1 Stability and Error Analysis for Kernel-Based Interpolation

As far as stability issues are concerned, it is well known that the ℓ_2 conditioning of a positive definite matrix K can be expressed as

$$cond(\mathsf{K}) = \frac{\lambda_{\max}}{\lambda_{\min}}, \tag{4.1}$$

where $\lambda_{\max}$ and $\lambda_{\min}$ are the largest and smallest eigenvalues of K, respectively.

Remark 4.1.1 *The definition of ℓ_2 conditioning in* (4.1) *is valid as long as we deal with a square normal matrix K, that is, it commutes with its conjugate transpose K^{H}. Such a definition is a particular case of the following formulation*

$$cond(\mathsf{A}) = \frac{\sigma_{\max}}{\sigma_{\min}},$$

which is true for any squared matrix A. Here, $\sigma_{\max}$ and $\sigma_{\min}$ are the largest and smallest singular values of A, respectively.

Concerning the largest eigenvalue, if K is the $N \times N$ positive definite kernel matrix related to a translational invariant kernel $\kappa(\boldsymbol{x},\boldsymbol{y}) = \phi(\boldsymbol{x}-\boldsymbol{y})$, an immediate consequence of Gershgorin's Theorem (see, e.g., Meyer, 2000) is that

$$\lambda_{\max} \leq N\phi(\mathbf{0}).$$

Later in this chapter, we present lower bounds for the smallest eigenvalue that are tailored to different kernels (see Diederichs and Iske, 2019; Schaback, 1994; and Fasshauer, 2007, section 16.1).

In order to provide the error analysis, we start by stating in Theorem 4.1.1 (Fasshauer, 2007, section 14.2).

Theorem 4.1.1 *Let $\Omega \subset \mathbb{R}^d$ and let κ be a strictly positive definite kernel. Then, for every set $X = \{\boldsymbol{x}_k,\ k = 1,\ldots,N\} \subset \Omega$ of distinct node points there exist functions $u_k^* \in span\{\kappa(\cdot,\boldsymbol{x}_k),\ k = 1,\ldots,N\}$ such that $u_i^*(\boldsymbol{x}_j) = \delta_{ij}$, where δ_{ij} is the well-known Kronecker delta.*

By means of the so-defined functions u_k^*, $k = 1,\ldots,N$, also called *cardinal functions*, we can write the interpolant in (2.2) of a given function $f : \Omega \longrightarrow \mathbb{R}$ in its *Lagrangian form* Wu and Schaback (1993), that is,

$$s(\boldsymbol{x}) = \sum_{k=1}^{N} f(\boldsymbol{x}_k)\, u_k^*(\boldsymbol{x}), \quad \boldsymbol{x} \in \Omega. \tag{4.2}$$

Let $\kappa \in \mathscr{C}(\Omega \times \Omega)$. For any set of distinct nodes X_N and vector $\boldsymbol{u} \in \mathbb{R}^N$, we can consider the quadratic form

$$Q(\boldsymbol{u}) := \kappa(\boldsymbol{x},\boldsymbol{x}) - 2\sum_{k=1}^{N} u_k \kappa(\boldsymbol{x},\boldsymbol{x}_k) + \sum_{i=1}^{N}\sum_{j=1}^{N} u_i u_j \kappa(\boldsymbol{x}_i,\boldsymbol{x}_j).$$

Then, we give Definition 4.1.1 (Fasshauer, 2007, section 14.3).

Definition 4.1.1 Let $\Omega \subseteq \mathbb{R}^d$ and let $\kappa \in C(\Omega \times \Omega)$ be a strictly positive definite kernel. Then, for any set of distinct points $X = \{\boldsymbol{x}_k,\ k = 1,\ldots,N\} \subseteq \Omega$, we define the *power function*

$$[P_{X,\kappa}(\boldsymbol{x})]^2 := Q(\boldsymbol{u}^*(\boldsymbol{x})),$$

where $\boldsymbol{u}^*$ is the vector of cardinal functions (see Theorem 4.1.1).

The power function can also be computed as

$$P_{X,\kappa}(\boldsymbol{x}) = \sqrt{\kappa(\boldsymbol{x},\boldsymbol{x}) - (\boldsymbol{b}(\boldsymbol{x}))^{\intercal}\mathsf{K}^{-1}\boldsymbol{b}(\boldsymbol{x})},$$

where $\boldsymbol{b} = (\kappa(\cdot,\boldsymbol{x}_1),\ldots,\ \kappa(\cdot,\boldsymbol{x}_N))^{\intercal}$ and K is the kernel matrix defined previously.

Moreover, letting K^y be the kernel matrix related to the augmented dataset $X \cup \{y\}$, we can also express the power function as (see De Marchi, 2003)

$$P_{X,\kappa}(\boldsymbol{y}) = \sqrt{\frac{\det \mathsf{K}^y}{\det \mathsf{K}}}. \tag{4.3}$$

Remark 4.1.2 *The power function vanishes if it is evaluated at the interpolation nodes, as emphasized by the formulation in* (4.3).

Thanks to the power function definition, we can now provide the following error bound for the kernel-based interpolant in (4.2) (Fasshauer, 2007, section 14.4).

Theorem 4.1.2 *Let $\Omega \subseteq \mathbb{R}^d$ and let $\kappa \in \mathscr{C}(\Omega \times \Omega)$ be a strictly positive definite kernel. Moreover, let $X = \{\boldsymbol{x}_k, k = 1, \ldots, N\} \subset \Omega$ be a set of distinct nodes. Then*

$$|f(\boldsymbol{x}) - s(\boldsymbol{x})| \leq P_{X,\kappa}(\boldsymbol{x}) \, \|f\|_{\mathscr{N}_\kappa(\Omega)}, \quad \boldsymbol{x} \in \Omega,$$

where $f \in \mathscr{N}_\kappa(\Omega)$.

Despite the dependency on the power function, and thus on the chosen set of nodes, we observe that Theorem 4.1.2 provides a pointwise error bound that does not give any indication concerning the *mesh size*, a term used in abuse of notation in the scattered data interpolation context. To obtain further details in this direction, we introduce the so-called *separation distance* and the *fill distance* (Fasshauer, 2007, section 14.1).

Definition 4.1.2 The separation distance is defined as

$$q_X := \frac{1}{2} \min_{i \neq j} \|\boldsymbol{x}_i - \boldsymbol{x}_j\|.$$

Definition 4.1.3 The fill distance is given by

$$h_{X,\Omega} := \sup_{\boldsymbol{x} \in \Omega} \min_{\boldsymbol{x}_k \in X} \|\boldsymbol{x} - \boldsymbol{x}_k\|.$$

Remark 4.1.3 *As a geometric intuition, the quantity q_X represents the radius of the largest ball that can be centered at every point in X so that there is no overlap. At the same time, $h_{X,\Omega}$ is the radius of the largest empty ball that can be placed among the data sites in Ω* (see Figure 4.1).

In what follows, we consider $h_{X,\Omega}$ to present a refined error bound for the kernel-based interpolant. In contrast, in Section 4.2, we consider q_X to study the stability of the interpolation procedure using different kernels.

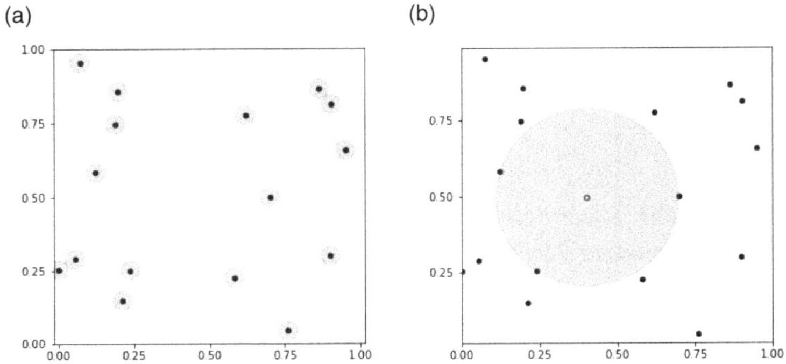

Figure 4.1 Given a dataset in $[0,1]^2$ (black dots), we show a representation of the separation distance (a) and of the fill distance (b), where the balls of interest are depicted in gray.

We introduce the following notation. Let $\boldsymbol{\beta} = (\beta_1, \ldots, \beta_M) \in \mathbb{N}_0^M$, with $|\boldsymbol{\beta}| = \sum_{k=1}^{M} \beta_k$. We define the differential operator $D^{\boldsymbol{\beta}}$ as

$$D^{\boldsymbol{\beta}} := \frac{\partial^{|\boldsymbol{\beta}|}}{(\partial x_1)^{\beta_1} \cdots (\partial x_M)^{\beta_M}}.$$

Moreover, we write $D_k^{\boldsymbol{\beta}} \kappa(\boldsymbol{x}, \boldsymbol{y})$ to indicate that the operator is applied to κ viewed as a function of the k-th variable.

Furthermore, we need to consider the following technical points.

Definition 4.1.4 The set $\Omega \subseteq \mathbb{R}^d$ satisfies an *interior cone condition* if there exist an angle $\theta \in (0, \pi/2)$ and a radius $\gamma > 0$ such that, for all $\boldsymbol{x} \in \Omega$, there exists a unit vector $\boldsymbol{\xi}(\boldsymbol{x})$ such that the cone

$$C = \{\boldsymbol{x} + \lambda \boldsymbol{y} : \boldsymbol{y} \in \mathbb{R}^d, ||\boldsymbol{y}||_2 = 1, \boldsymbol{y}^\intercal \boldsymbol{\xi}(\boldsymbol{x}) \geq cos(\theta), \lambda \in [0, \gamma]\}$$

is contained in Ω.

Finally, we can state the following error bound in terms of the fill distance $h_{X,\Omega}$ (Fasshauer, 2007, section 14.5).

Theorem 4.1.3 *Let $\Omega \subseteq \mathbb{R}^d$ be bounded and suppose that it satisfies an interior cone condition. Moreover, let $\kappa \in \mathscr{C}^{2k}(\Omega \times \Omega)$ be a strictly positive definite kernel. Then, there exist positive constants h_0 and C, independent of $\boldsymbol{x}$, f, and κ, such that*

$$|f(\boldsymbol{x}) - s(\boldsymbol{x})| \leq C h_{X,\Omega}^k \sqrt{C_\kappa(\boldsymbol{x})} ||f||_{\mathscr{N}_\kappa(\Omega)},$$

provided $h_{X,\Omega} \leq h_0$ *and* $f \in \mathcal{N}_\kappa(\Omega)$*, where*

$$C_\kappa(\boldsymbol{x}) = \max_{|\boldsymbol{\beta}|=2k} \left(\max_{\boldsymbol{w},\boldsymbol{z} \in \Omega \cap B(\boldsymbol{x}, c_2 h_{X,\Omega})} \left| D_2^{\boldsymbol{\beta}} \kappa(\boldsymbol{w}, \boldsymbol{z}) \right| \right),$$

with $B(\boldsymbol{x}, c_2 h_{X,\Omega})$ *denoting the ball of radius* $c_2 h_{X,\Omega}$ *centered at* $\boldsymbol{x}$*, where* c_2 *is derived from* Fasshauer (2007, Theorem 14.4).

Remark 4.1.4 *A consequence of Theorem 4.1.3 is that the approximation order* k *is controlled by the smoothness of the kernel function. Moreover, if* $k \in \mathbb{N}$ *and* $|\boldsymbol{\alpha}| \leq k$*, we have*

$$|D^{\boldsymbol{\alpha}} f(\boldsymbol{x}) - D^{\boldsymbol{\alpha}} s(\boldsymbol{x})| \leq C_k h_{X,\Omega}^{k-|\boldsymbol{\alpha}|} \|f\|_{\mathcal{N}_\kappa(\Omega)},$$

whenever $f \in \mathcal{N}_\kappa(\Omega)$.

Further estimates tailored for kernels of different smoothness are provided in the following sections. To such an aim, we need to point out that radial kernels usually include a real shape parameter $\varepsilon > 0$ in their definition, that is,

$$\kappa(\boldsymbol{x}, \boldsymbol{y}) = \varphi(r) \Longrightarrow \varphi(\varepsilon r) = \varphi_\varepsilon(r) = \kappa_\varepsilon(\boldsymbol{x}, \boldsymbol{y}), \quad r = \|\boldsymbol{x} - \boldsymbol{y}\|.$$

4.2 Examples of Strictly Positive Definite Radial Functions

Here, we present details concerning some strictly positive definite radial functions, which guarantee the existence of a unique solution in Problem 2.1.1. We refer to Fasshauer (2007, chapters 4, 15, and 16) for more details and functions.

4.2.1 The Gaussian Function

We consider the well-known Gaussian function denoted as

$$\varphi_G(r) = e^{-(\varepsilon r)^2}. \tag{4.4}$$

The function φ_G is $\mathscr{C}^\infty(\mathbb{R})$ and globally supported. It is widely used and employed in different areas, such as solving the heat and diffusion equations in the theory of partial differential equations (PDEs) and in statistics, in order to describe the normal distribution.

We can easily write the d-dimensional Fourier transform of $\phi_G(\boldsymbol{x}) = \varphi_G(\|\boldsymbol{x}\|)$, that is,

$$(\mathrm{F}\phi_G)(\boldsymbol{\omega}) = \frac{1}{(\sqrt{2}\varepsilon)^M} e^{-\|\boldsymbol{\omega}\|^2/(4\varepsilon^2)}.$$

Since the Fourier transform of a Gaussian is again a Gaussian, it is strictly positive definite (see Theorem 2.2.4).
The native space of Gaussians contains the class of *band-limited* functions, which are functions whose Fourier transform is compactly supported.

As a consequence of Theorem 15.1 in Fasshauer (2007), the Gaussian kernel yields a *spectral* order of approximation for the interpolation of functions within its associated native space, since this kernel is infinitely smooth.

On the other hand, concerning the stability of the interpolation procedure, we have the following lower bound for the smallest eigenvalue of the kernel matrix,

$$\lambda_{\min} \geq C_d(\sqrt{2}\varepsilon)^{-d} e^{-40.71d^2/(q_X\varepsilon)^2} q_X^{-d}. \tag{4.5}$$

We observe that the lower bound approaches zero exponentially as the separation distance decreases, leading to exponential growth in the condition numbers of the kernel matrices and potentially causing instability in the interpolation process.

Example 4.2.1 (Gaussian functions) In Figure 4.2, *we display two examples of Gaussian functions in* $[-1,1]^2$*, focusing on the role played by the shape parameter.*

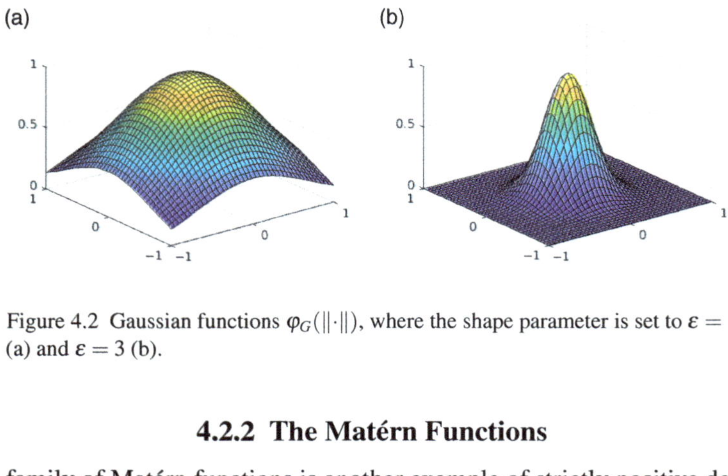

Figure 4.2 Gaussian functions $\varphi_G(\|\cdot\|)$, where the shape parameter is set to $\varepsilon = 1$ (a) and $\varepsilon = 3$ (b).

4.2.2 The Matérn Functions

The family of Matérn functions is another example of strictly positive definite functions with global support. As Gaussian, they are also commonly used in statistics Matérn (1986).
More precisely, by taking $\beta \in \mathbb{R}$, $\beta > d/2$, the Matérn functions are defined as

$$\phi_M^\beta(\boldsymbol{x}) = \frac{J^2_{\beta-d/2}(\|\boldsymbol{x}\|)\|\boldsymbol{x}\|^{\beta-d/2}}{2^{\beta-1}\Gamma(\beta)}, \tag{4.6}$$

where $J^2_{\beta-d/2}$ is the *modified Bessel function of the second kind* of order $\beta - d/2$. The Fourier transform of such functions is given by the *Bessel kernels*

$$(\mathrm{F}\phi_M^{\beta})(\boldsymbol{\omega}) = (1+\|\boldsymbol{\omega}\|^2)^{-\beta}, \tag{4.7}$$

thus they are strictly positive definite on $\mathbb{R}^d$ for $\beta > d/2$.

We point out that the native space related to the Matérn function ϕ_M^{β} is the Sobolev space

$$\mathscr{W}_2^{\beta}(\mathbb{R}^d) = \{f \in \mathscr{L}_2(\mathbb{R}^d) \cap \mathscr{C}(\mathbb{R}^d) \,:\, (\mathrm{F}f)(\cdot)(1+\|\cdot\|^2)^{\beta/2} \in \mathscr{L}_2(\mathbb{R}^d)\}.$$

This is the reason why Matérn functions are often referred to as *Sobolev splines*. Moreover, we have the following error bound

$$|D^{\boldsymbol{\alpha}} f(\boldsymbol{x}) - D^{\boldsymbol{\alpha}} r(\boldsymbol{x})| \le C h_{X,\Omega}^{\beta-d/2-|\boldsymbol{\alpha}|} \|f\|_{\mathscr{N}_\kappa(\Omega)}, \tag{4.8}$$

provided $|\boldsymbol{\alpha}| \le \beta - \lceil \frac{d+1}{2} \rceil$ and f in the native space $\mathscr{N}_\kappa(\Omega)$.

By considering functions of limited regularity, such as the Matérn functions, it is possible to adapt the chosen radial function to the data at hand, possibly preventing instability issues linked to smooth kernels (as described in 4.5).

Example 4.2.2 (Matérn functions) *We present the following Matérn functions, along with their regularity degrees, namely*

$$\begin{array}{lllll} \varphi_{M,0}(r) & = & e^{-\varepsilon r}, & \beta = \frac{d+1}{2} & C^0 \\ \varphi_{M,2}(r) & = & e^{-\varepsilon r}(1+\varepsilon r), & \beta = \frac{d+3}{2} & C^2 \\ \varphi_{M,4}(r) & = & e^{-\varepsilon r}(3+3\varepsilon r+(\varepsilon r)^2), & \beta = \frac{d+5}{2} & C^4. \end{array} \tag{4.9}$$

In Figure 4.3*, we display two examples of such functions in* $[-1,1]^2$.

4.2.3 The Wendland Functions

The kernel matrices generated by globally defined kernels are, of course, dense, and to partially overcome this issue, one may be interested in studying compactly supported kernels. One of the most important families of compactly supported radial functions is the Wendland kernels. To introduce them, we need to define the following operators (Schaback and Wu, 1996).

Definition 4.2.1 Let φ be such that $r \to r\varphi(r) \in \mathscr{L}_1[0,\infty)$. Then the integral operator I is given by

$$(\mathrm{I}\varphi)(r) = \int_r^{\infty} r\phi(r)\mathrm{d}t, \quad r \ge 0.$$

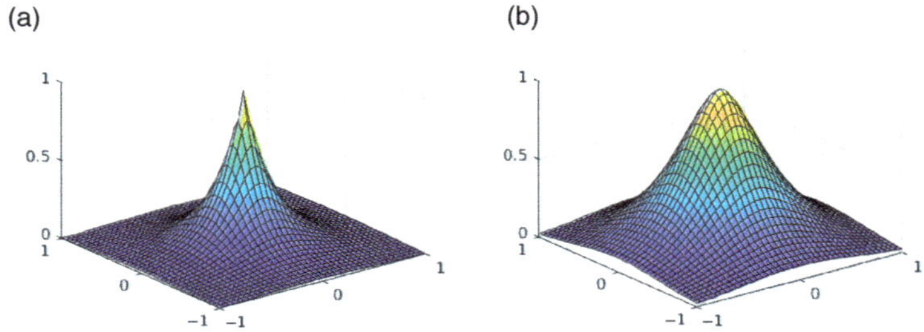

Figure 4.3 Functions (a) $\varphi_{M,0}(\|\cdot\|)$ and (b) $\varphi_{M,2}(\|\cdot\|)$ are represented. The shape parameter is set to $\varepsilon = 4$.

Definition 4.2.2 Let $\varphi \in \mathscr{C}^2(\mathbb{R})$ be an even function. The differential operator D is defined as

$$(\mathrm{D}\varphi)(r) = -\frac{1}{r}\varphi'(r), \quad r \geq 0.$$

We present some properties of these operators and their important consequences, which are useful for defining new classes of radial functions.

Proposition 4.2.1 *Some properties of the operators* I *and* D *are as follows:*

(i) If φ has compact support, then so do $\mathrm{D}\varphi$ *and* $\mathrm{D}\varphi$.
(ii) If $\varphi \in \mathscr{C}(\mathbb{R})$ and $r \to r\varphi(r) \in \mathscr{L}_1[0,\infty)$, then $\mathrm{DI}\varphi = \varphi$.
(iii) If $\varphi \in \mathscr{C}^2(\mathbb{R})$ is even and not identically equal to one, and $\varphi' \in \mathscr{L}_1[0,\infty)$, then $\mathrm{ID}\varphi = \varphi$.
(iv) If $r \to r^{d-1}\varphi(r) \in \mathscr{L}_1[0,\infty)$ and $d \geq 3$, then the Hankel transform of order d of φ is equal to the Hankel transform of order $d-2$ of $\mathrm{I}\varphi$.
(v) If $\varphi \in \mathscr{C}^2(\mathbb{R})$ is even and $r \to r^d\varphi'(r) \in \mathscr{L}_1[0,\infty)$, then the Hankel transform of order d of φ is equal to the Hankel transform of order $d+2$ of $\mathrm{D}\varphi$.

Recalling the results in Theorem 2.2.5, we observe that operators I and D relate d-dimensional Fourier transforms to $(d-2)$- or $(d+2)$-dimensional ones. Theorems 4.2.1–4.2.3 are consequences of the properties presented.

Theorem 4.2.1 *Let $\varphi \in \mathscr{C}(\mathbb{R})$, $r \to r^{d-1}\varphi(r) \in \mathscr{L}_1[0,\infty)$, and $d \geq 3$. Then, $\varphi(\|\cdot\|)$ is strictly positive definite and radial on $\mathbb{R}^d$ if and only if $\mathrm{I}\varphi(\|\cdot\|)$ is strictly positive definite and radial on $\mathbb{R}^{d-2}$.*

Theorem 4.2.2 *Let $\varphi \in \mathscr{C}^2(\mathbb{R})$ be even and let $r \to r^d\varphi'(r) \in \mathscr{L}_1[0,\infty)$. Then,*

$\varphi(\|\cdot\|)$ is strictly positive definite and radial on $\mathbb{R}^d$ if and only if $\mathrm{D}\varphi$ is strictly positive definite and radial on $\mathbb{R}^{d+2}$.

The family of Wendland functions, constructed in Wendland (1995), consists of functions that are compactly supported, strictly positive definite, and smooth. To define this family of functions, we consider the *truncated power function* $\varphi_{T,k}(r) = (1-r)_+^k$. We point out that $\varphi_{T,k}(\|\cdot\|)$ is strictly positive definite and radial on $\mathbb{R}^d$ for $k \geq \lfloor d/2 \rfloor + 1$.
Then, the Wendland functions are constructed repeatedly applying the operator I, that is,

$$\varphi_W^{d,k} = \mathrm{I}^k \varphi_{T,\lfloor d/2 \rfloor + k + 1}.$$

Theorem 4.2.3 *The functions $\varphi_W^{d,k}(\|\cdot\|)$ are strictly positive definite and radial on $\mathbb{R}^d$. More precisely, they assume the form*

$$\varphi_W^{d,k}(r) = \begin{cases} p^{d,k}(r), & r \in [0,1], \\ 0, & r > 1, \end{cases}$$

with $p^{d,k}$ being a univariate polynomial of degree $\lfloor d/2 \rfloor + 3k + 1$. Moreover, $\varphi_W^{d,k} \in \mathscr{C}^{2k}(\mathbb{R})$ are unique up to a constant factor, and the polynomial degree is minimal for given space dimension d and smoothness $2k$.

Similar to the Matérn functions, the native space of the Wendland function $\varphi_W^{d,k}(\cdot) = \phi_W^{d,k}(\|\cdot\|)$ is the Sobolev space $\mathscr{W}_2^{d/2+k+1/2}(\mathbb{R}^d)$, with the restriction $d \geq 3$ in the case $k = 0$. Furthermore, we have the following error bound

$$|D^{\boldsymbol{\alpha}} f(\boldsymbol{x}) - D^{\boldsymbol{\alpha}} r(\boldsymbol{x})| \leq C h_{X,\Omega}^{k+1/2-|\boldsymbol{\alpha}|} \|f\|_{\mathscr{N}_\kappa(\Omega)}$$

provided $|\boldsymbol{\alpha}| \leq k$, $h_{X,\Omega}$ sufficiently small and f in the native space $\mathscr{N}_\kappa(\Omega)$. Concerning the stability of the interpolation process, the smallest eigenvalue of the kernel matrix is bounded from above by $\lambda_{\min} \geq C_{d,k} q_X^{2k+1}$, where $C_{d,k}$ is a constant depending on d and k.

Therefore, the lower bound goes to zero with q_X at a polynomial rate, providing a better conditioning with respect to the Gaussian case (see (4.5)).

Example 4.2.3 (Wendland functions) *We report some well-known examples of Wendland functions (with $d = 3$) that we will consider, along with their regularity degrees. They are strictly positive definite and radial on $\mathbb{R}^s$ for $s \leq 3$.*

$$\begin{aligned} \varphi_{W,0}(r) &\doteq (1-\varepsilon r)_+^2, && C^0 \\ \varphi_{W,2}(r) &\doteq (1-\varepsilon r)_+^4 (4\varepsilon r + 1), && C^2 \\ \varphi_{W,4}(r) &\doteq (1-\varepsilon r)_+^6 \left(35(\varepsilon r)^2 + 18\varepsilon r + 3\right), && C^4 \end{aligned} \tag{4.10}$$

where $\doteq$ denotes equality up to a positive constant factor. In our applications, we set the constant factor to 1, thereby imposing equalities in the definition. In Figure 4.4, *we display two examples in* $[-1,1]^2$.

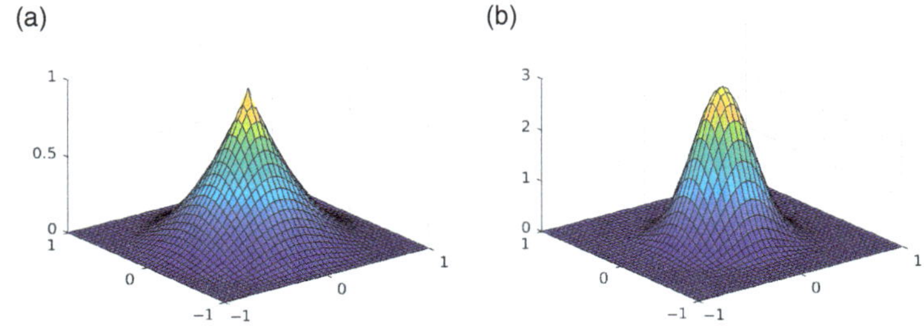

Figure 4.4 Functions (a) $\varphi_{W,0}(\|\cdot\|)$ and (b) $\varphi_{W,2}(\|\cdot\|)$ are displayed. The shape parameter is set to $\varepsilon = 1$.

4.3 Examples of Completely and Multiply Monotone Functions

Here, we enumerate some of the most common Completely Monotone (CM) functions, and some facts according to Section 2.2.3.

1. The function $\varphi(r) = \varepsilon$, $\varepsilon \geq 0$ is CM on $[0,\infty)$.
2. The function $\varphi(r) = e^{-\varepsilon r}$, $\varepsilon \geq 0$ is CM on $[0,\infty)$, indeed
$$(-1)^k \varphi^{(k)}(r) = \varepsilon^k e^{-\varepsilon r} \geq 0, \quad k = 0,1,2,...$$
By evaluating at r^2, we obtain the family of Gaussians.
3. The function $\varphi(r) = \frac{1}{(1+\varepsilon r)^\beta}$, $\beta \geq 0$ is CM on $[0,\infty)$, indeed
$$(-1)^k \varphi^{(k)}(r) = (-1)^{2k} \beta(\beta-1)\cdots(\beta+k-1)(1+\varepsilon r)^{-\beta-k} \geq 0,$$
for $k = 0,1,2,\ldots$. By evaluating at r^2, we obtain the family of *inverse multiquadrics*, a family of $\mathscr{C}^\infty$ strictly positive definite functions.

The *truncated power* (see Figure 4.5)

$$\varphi_{T,k}(r) = (1-\varepsilon r)_+^k$$

is k-times monotone for any k. Indeed

$$(-1)^\beta \varphi_{T,k}^{(\beta)} = k(k-1)\cdots(k-\beta+1)(1-r)_+^{k-\beta} \geq 0, \ \beta = 0,1,\ldots,k.$$

In Section 4.2, these functions were used to construct strictly positive-definite Wendland radial functions.

Example 4.3.1 (Truncated power functions) In Figure 4.5, *we display* $\varphi_{T,k}$ *for* $k = 1,4$, *in* $[-1,1]^2$.

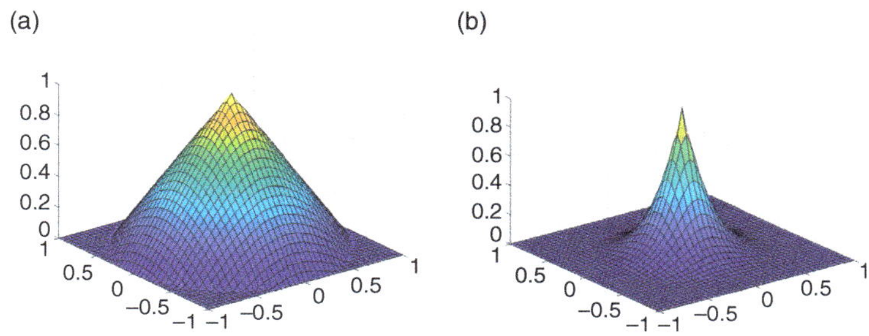

Figure 4.5 Functions (a) $\varphi_{T,1}(\|\cdot\|)$ and (b) $\varphi_{T,4}(\|\cdot\|)$ are displayed. The shape parameter is set to $\varepsilon = 1$.

4.4 Examples of Conditionally Positive Definite Functions

In this section, we list some conditionally positive definite (CPD) radial functions and their properties.

The *generalized multiquadrics* are defined as

$$\varphi(r) = (1+(\varepsilon r)^2)^{\beta}, \quad \beta \in \mathbb{R}\backslash\mathbb{N}_0.$$

They are CPD of order $m = \lceil \beta \rceil$, and $\mathscr{C}^{\infty}$ at the origin. In the definition, positive integer values for β are excluded to avoid polynomials of *even* degree. For $\beta = 1/2$, we obtain the well-known *Hardy multiquadric* that is CPD of order 1.

The following are particular *shape-parameter-free* RBFs:

1. *Radial powers*

$$\varphi_{R,\beta}(r) = r^{\beta}, \ , \ 0 < \beta \notin 2\mathbb{N}$$

are CPD of order $m = \lceil \beta/2 \rceil$ (and higher). For $\beta = 3$, we get a function that is CPD of order 2 and for $\beta = 5$ a function that is CPD of order 3 (see Figure 4.6).

2. *Thin-plate splines*

$$\varphi_{S,\beta}(r) = r^{2\beta}\log r, \ \ \beta \in \mathbb{N}$$

are CPD of order $m = \beta + 1$ (see Figure 4.7).

Example 4.4.1 (Radial powers) In Figure 4.6, *we display* $\varphi_{R,\beta}$ *for* $\beta = 3, 5$, *in* $[-1, 1]^2$.

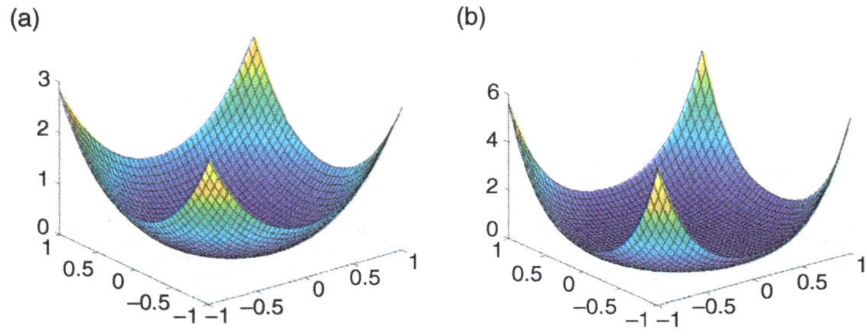

Figure 4.6 Functions (a) $\varphi_{R,3}$ and (b) $\varphi_{R,5}$ are displayed.

Example 4.4.2 (Thin-plate splines) In Figure 4.7, *we display* $\varphi_{S,\beta}$ *for* $\beta = 1, 2$, *in* $[-1, 1]^2$.

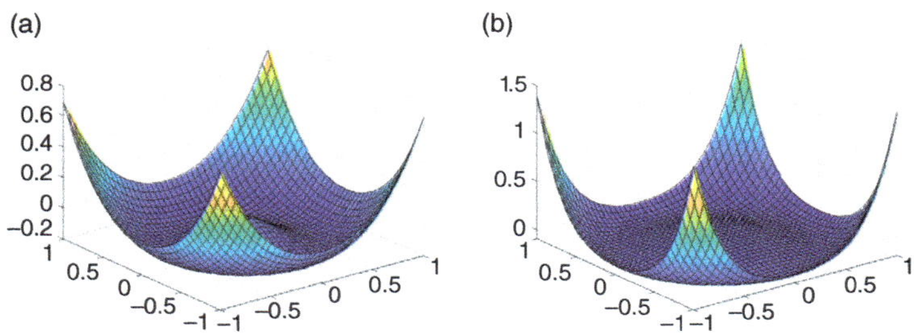

Figure 4.7 Functions (a) $\varphi_{S,1}$ and (b) $\varphi_{S,2}$ are displayed.

5

On Uncertainty Principles and Shape Parameter Selection

The shape parameter in kernel functions affects both accuracy and numerical stability, leading to fundamental uncertainty principles. We recall three main trade-offs: accuracy versus stability, accuracy and stability versus problem size, and accuracy versus efficiency. We present empirical strategies for shape parameter tuning, including power function minimization and cross-validation (CV) techniques such as the extended Rippa's algorithm. These methods enable practical, data-driven parameter selection without requiring additional knowledge of the underlying function.

5.1 RBF Approximation and Uncertainty Principles

Computationally, we might observe some incongruence with the previously presented theory. These issues are known as the trade-off principle within the scientific community and are briefly discussed in the following sections.

1. **Accuracy VS stability**

   The concept of condition number is related to the first *uncertainty principle* Schaback (1995), which consists of a conflict between theoretical accuracy and numerical stability.

   - If a large number of interpolation nodes are involved, the interpolation systems may suffer from ill-conditioning.
   - The conditioning is linked to the order of the basis functions and to the node distribution.
   - The ill-conditioning increases mainly due to the decrease in the separation distance q_X and not necessarily due to the increase in data points.
   - If we reduce the value of the shape parameter, then the condition number

increases. This is consistent with the fact that, for small values of ε, the basis functions are very flat, resembling constant functions.

2. **Accuracy and stability VS problem size**
   The second trade-off principle concerns accuracy and stability versus problem size. This is perhaps the most studied computational problem. Thus, in what follows, we point out some important facts, and we give several references
   - The problem of stably computing generalized inverse multiquadrics and Gaussian interpolants, with $\varepsilon \longrightarrow 0$, has been investigated in many papers, see, for example, Driscoll and Fornberg (2002) and Fornberg and Wright (2004).
   - More recently, considering a wider family of kernels, several authors proposed new techniques that allow stable interpolation for extreme values of the shape parameter and with relatively low computational complexity (De Marchi and Santin, 2013, 2015).
3. **Accuracy VS efficiency**
   Of course, this trade-off involves kernels with compact support, such as the Wendland functions. In particular:
   - If we scale the support size of the basis functions proportionally to the fill distance, the resulting method is numerically stable, but there is essentially no convergence. This is the *stationary* approximation .
   - In the *nonstationary* approximation, that is, with fixed support size, the error decreases, but the interpolation matrices become dense and thus ill-conditioned.

In general, for many choices of basis functions, the kernel systems are guaranteed to be non-singular whenever the nodes are distinct. Problems arise anyway if the separation distance is *small* as the condition number grows. Then, in agreement with the trade-off principles (Fasshauer, 2007; Schaback, 1995), for a low data point density, high-order basis functions, such as Gaussians, are strongly recommended. On the other hand, for a high density of interpolation points, we can use low-order basis functions, such as compactly supported Wendland or Matérn functions.

Remark 5.1.1 *We point out that these trade-off or uncertainty principles have been questioned in Fasshauer and McCourt (2015, chapter 11), where the authors observe that stability and effectiveness can be achieved by departing from standard kernel bases and adopting stabilisation strategies. Nevertheless, trade-off principles still represent a helpful guide, especially when one is not entirely familiar with kernel interpolation.*

5.2 Empirical Error Estimates and Shape Parameter Tuning

The shape parameter in the kernel definition provides some flexibility in defining the shape of the underlying RBFs. Indeed, increasing this parameter produces more localized basis functions. In contrast, the decrease of such a parameter determines "flat" basis functions (see, e.g., Figure 4.2).

Different shape parameters induce different native spaces; consequently, the accuracy of the kernel-based interpolant varies with the choice of the parameter. Therefore, after choosing the RBF, error estimation and shape parameter tuning are strongly intertwined.

It turns out that the tuning of the shape parameter is a nontrivial issue in RBF interpolation. First, in real applications, we usually do not know the underlying function that generates the interpolation data; thus, the choice of the shape parameter should depend solely on the data at our disposal, without reference to the ground truth. Moreover, the tuning of this parameter is typically an unstable process: slight variations in the shape parameter around the optimal value could worsen the interpolation error by orders of magnitude (Fasshauer, 2007, section 2.2). We remark that a *trial-and-error* approach, performed through various interpolation experiments with different radial bases and values of the shape parameter, can be carried out in some situations (if ground truth is available).

Therefore, many different strategies for tuning the shape parameter have been proposed and studied in recent years. Here, we focus our presentation on two approaches that will also be considered later on in Sections 5.2.1 and 5.2.2.

5.2.1 Using the Power Function

As a consequence of Theorem 4.1.2, the shape parameter can be chosen by evaluating the maximum of the power function on the whole domain (in a trial and error version) or at some further data sites at disposal (in a more applied context) (Fasshauer, 2007, section 17.1). The limitation of this approach is to forget the second part of the error estimate, that is, the one that depends on the basis function via the native space norm of f. On the other hand, since the power function does not depend on the particular underlying function to be approximated, what is obtained via this approach can be quite effective for many targets but is still sub-optimal.

5.2.2 Using Cross-validation with the Extended Rippa's Algorithm

Other algorithms that take into account the residuals are CV schemes (Golub et al., 1979), which are an effective statistical technique widely applied in many fields, including the machine learning classification setting we introduce in Chapter 12. Therefore, we present a CV in a more general setting that is not necessarily associated with kernel-based interpolation.

5.2.2.1 General Setting

Let $\Omega \subset \mathbb{R}^d$ and let $X = \{\boldsymbol{x}_k,\ k = 1,\dots,N\} \subset \Omega$ be a set of data, where $N,d \in \mathbb{N}$ and $N,d \geq 1$. We can associate with X a set of labels Y, whose nature depends on the considered task (e.g., we can have data values $Y \subset \mathbb{R}$ considering regression or interpolation tasks, or classes represented by integers $Y \subset \mathbb{Z}$ in classification problems). Then, we can define the labeled dataset

$$Z = \{(\boldsymbol{x}_k, y_k),\ \boldsymbol{x}_k \in X,\ y_k \in Y\} \subset X \times Y.$$

In principle, let us suppose that we aim to describe and analyze a process or task involving our data, and to make predictions about possible new instances. Letting $\Theta \subseteq \mathbb{R}^t$, $t \in \mathbb{N}$, be a hyperparameter space, we can consider a set of models $\{m_{\boldsymbol{\theta}}\}_{\boldsymbol{\theta}\in\Theta}$, where every model depends on a vector of real parameters $\boldsymbol{\theta} = (\theta_1,\dots,\theta_t)$.

The role of CV is to identify the model that better represents the known data. In doing this, CV iteratively considers a part of the data at its disposal as unknown and constructs (or trains) the model with the remaining data. The aim is to test the generalization capability of each model, that is, how the model performs with *new* unknown test samples.

Fixed $\boldsymbol{\theta} \in \Theta$, the CV scheme proceeds as follows.

(i) Let $k \in \mathbb{N}$, $1 < M \leq N$. We divide the set Z into k (possibly equal sized) disjoint sets $\{Z^i\}_{i=1,\dots,k}$, which are usually called *folds*.
(ii) We choose $k-1$ *training* folds $T_1,\dots,T_{k-1}$ out of Z, denoting with V the remaining *validation* fold, and train the model $m_{\boldsymbol{\theta}}$ on $T = \bigcup_{i=1}^{k-1} T_i$.
(iii) We take a cost function (e.g., the zero-one loss in classification tasks or the residual in regression settings) and evaluate the model by computing the cost on V.
(iv) This process is repeated k times, choosing different training and validation fold in Z. Finally, we average the costs obtained during the entire process, then assign the average cost to the model $m_{\boldsymbol{\theta}}$.

Varying then $\boldsymbol{\theta} \in \Theta$, we choose the model that realises the minimum average cost. This procedure is also known as k-fold CV. Moreover, in the classification setting, we refer to *stratified* CV when the folds are determined in such a way that the proportions between the labels in Y are preserved in every fold. Finally, we refer to *nested* CV where, given the CV validation setting, another nested CV scheme is constructed on each training fold.

5.2.2.2 The Extended Rippa's Algorithm

As noted previously, in RBF interpolation, the interpolant s (i.e., the model) is usually built on a kernel κ_ε that depends on a positive shape parameter ε. Therefore, to calibrate the model, different values of the shape parameter $\varepsilon_1, \ldots, \varepsilon_t$ must be evaluated (i.e., $\boldsymbol{\theta} = (\varepsilon_1, \ldots, \varepsilon_t)$). For this purpose, a novel approach based on machine learning techniques is detailed in Chapter 15.

Suppose for simplicity that $p = N/k \in \mathbb{N}$. For each value of the shape parameter, a standard application of the k-fold CV scheme would require the inversion of k different $(N-p) \times (N-p)$ linear systems, leading to a complexity cost of approximately $\mathcal{O}(N^3 k)$, which approaches $\mathcal{O}(N^4)$ as the number of folds increases. An efficient strategy to reduce this computational cost, known as the extended Rippa's algorithm (ERA), has been presented recently in Marchetti (2021). To go into details, we introduce some further notation in the following lines.

For each fold of the CV scheme, we split the dataset X into a training set T_{N-p} of cardinality $N-p$ and validation set V_p, so that $X = T_{N-p} \cup V_p$ and $T_{N-p} \cap V_p = \emptyset$. Let us denote by $\boldsymbol{v} = (v_1, \ldots, v_p)^\mathsf{T}$, $v_i \in \{1, \ldots, N\}$ the vector of distinct validation indices for a given fold, that is, $V_p = \{\boldsymbol{x}_{v_i}, i = 1, \ldots, p\}$. Then, we are interested in computing the residual vector $\boldsymbol{r} = (r_1, \ldots, r_p)$, whose components are $r_i = f(\boldsymbol{x}_{v_i}) - s_{T_{N-p}}(\boldsymbol{x}_{v_i})$, $\boldsymbol{x}_{v_i} \in V_p$, where $s_{T_{N-p}}$ is the interpolant constructed on the training set T_{N-p}. In Marchetti (2021), the author proves that $\boldsymbol{r}$ is the unique solution of the linear system

$$\mathsf{K}^{-1}_{\boldsymbol{v},\boldsymbol{v}} \boldsymbol{w} = \boldsymbol{c}_{\boldsymbol{v}},$$

where $\mathsf{K}^{-1}_{\boldsymbol{v},\boldsymbol{v}}$ is the submatrix of the inverse kernel matrix K^{-1} which is defined by restricting to the rows and columns of the validation indices, and $\boldsymbol{c}_{\boldsymbol{v}}$ are the components of the solution vector of (2.3) for which $i \in \boldsymbol{v}$. Hence, such an implementation requires a total complexity cost of about $\mathcal{O}(N^3) + \mathcal{O}(N^3/k^2)$.

Remark 5.2.1 *The ERA is indeed the extension of a well-known algorithm introduced in Rippa (1999) for Leave-One-Out CV (LOOCV), i.e., the case $k = N$. Indeed, if $\boldsymbol{v} = v \in \{1, \ldots, N\}$ then*

$$\boldsymbol{r}_v = \frac{\boldsymbol{c}_v}{\mathsf{K}^{-1}_{v,v}}.$$

We note, that in Chapters 8 and 15, we will also use the ERA extension for slightly different purposes.

Many other techniques are available for the tuning of the shape parameter. *Maximum Likelihood Estimation (MLE)* is a statistical technique that has both advantages and disadvantages compared to CV Neumaier (1998). Modifications of Rippa's leave-one-out CV (LOOCV) scheme have been presented in Fasshauer and Zhang (2007) and Trahan and Wyatt (2003), and applied beyond the approximation setting in Cavoretto et al. (2024); Cavoretto and De Rossi (2020); Cavoretto et al. (2018); and Chen et al. (2020), while in Ling and Marchetti (2022) a modified version of the ERA has been proposed, which computes an approximation of the exact CV scheme. A further generalization to the generalized interpolation framework has been studied in Marchetti (2024). Probabilistic approaches for setting the shape parameter value have been considered in Chiu et al. (2020) and Yang et al. (2018). For the Gaussian kernel, the so-called RBF-QR method is particularly effective in Fornberg and Piret (2008) for the sphere, and Fornberg, Larsson and Flyer (2011) for 1D, 2D and 3D. Finally, for multiquadrics and Gaussians up to two-dimensional, a Taylor-based criterion for the optimal value of the shape parameter has been derived in Bayona et al. (2012). Finally, see Sun and Wang (2026) for a recent overview on RBF parameters' tuning approaches.

6
Rational Kernels

In this chapter, we focus on improving the computational stability of kernel-based approximation methods by modifying the underlying kernel basis. In particular, we discuss recent research directions related to rational RBF approximation. The approach presented here, known as the eigen-rational kernel-based scheme, is built on a fractional RBF expansion, where the denominator is defined by the eigenvector associated with the largest eigenvalue of the kernel matrix. Kernel-generated spaces are well known for providing strong theoretical and practical advantages in multivariate approximation. However, in certain applications, it can be beneficial to move beyond linear approximation spaces and consider nonlinear spaces generated by RBFs. Rational RBF approximants represent one such extension. A key drawback of standard rational RBF schemes is that the rational basis depends explicitly on function values rather than solely on the distribution of data sites. The eigen-rational approach overcomes this limitation by constructing the rational basis from spectral information in the kernel matrix, thereby improving numerical stability.

6.1 Overview of Stable Computation Techniques

One of the most investigated topics in this field is the stable computation of the kernel interpolant, also in the *flat* limit, that is, when $\varepsilon \longrightarrow 0$.

In agreement with the trade-off principles, we remark the following:

- In some cases, the approximants may suffer from instability because of ill-conditioning of the interpolation matrices.
- This problem depends on both the smoothness order of the basis function and the node distribution. More specifically, if one keeps the number of nodes

fixed and considers smooth basis functions, then the problem of instability becomes evident for small values of the shape parameter.

- Of course, a basis function with a finite order of smoothness can be used to improve the conditioning, but the accuracy of the fit gets worse. For this reason, recent research has moved toward the study of stable bases.

Among the most popular methods for stabilizing the interpolant, many belong to the following categories:

1. **RBF–QR Method**: It is rooted in a particular decomposition of the kernel, and it has been developed so far to treat the Gaussian kernel; refer, for example, to Fornberg and Piret (2008) for the sphere.
2. **Hilbert–Schmidt singular value decomposition (HS–SVD)**: It has been developed to stably compute the RBF interpolants (Cavoretto et al., 2015; Fasshauer and McCourt, 2015). In principle, this technique can be applied to any kernel, provided that the HS eigenvalues and eigenvectors are known. However, these quantities are far from being easy to compute, and in practice, they only work for the Gaussian function.
3. **Weighted SVD (WSVD) Bases**: It is a more general approach that applies to any RBF, consisting of computing a WSVD decomposition which produces stable bases; see De Marchi and Santin (2013). This method will be introduced and analyzed in the sequel.

In this chapter, as in the methods discussed previously, we aim to change the kernel basis to improve its computational stability. We propose some recent research topics on rational RBF approximation. In particular, we review the main theoretical and computational aspects studied in Buhmann et al. (2020). The method here reported is the so-called eigen-rational kernel-based scheme. It consists of a fractional RBF expansion, with the denominator depending on the eigenvector associated to the largest eigenvalue of the kernel matrix.

We have already seen that the spaces spanned by kernels provide beneficial properties for multivariate approximation. Nevertheless, it might be advantageous to study approximants in nonlinear spaces generated by RBFs. Few examples of these approaches already exist in the literature and are known as rational RBFs, introduced in Jakobsson et al. (2009) and further developed in De Marchi et al. (2019) and Sarra and Bay (2018). The main disadvantage of the scheme proposed in these papers is that the *rational basis* is constructed from function values, and consequently it depends not only on data sites. A solution comes from the eigen-rational basis.

6.2 Eigen-Rational Kernel-Based Interpolation

This study is motivated by the analogy with a polynomial rational approximation, which is well known to perform particularly well. However, it is a mesh-dependent approach. The class of kernel-based rational interpolants proposed in Buhmann et al. (2020), namely eigen-rational interpolants, assumes the form

$$\hat{s}_f(\boldsymbol{x}) = \frac{\sum_{i=1}^{N} c_i \kappa(\boldsymbol{x}, \boldsymbol{x_i}) + \sum_{m=1}^{M} d_m p_m(\boldsymbol{x})}{\sum_{k=1}^{N} b_k \bar{\kappa}(\boldsymbol{x}, \boldsymbol{x_k})} = \frac{s_g(\boldsymbol{x})}{s_h(\boldsymbol{x})},$$

defined for some function values g_i, h_i, $i = 1, \ldots, N$, where κ is conditionally positive definite, while $\bar{\kappa}$ is a strictly positive definite kernel.

Remark 6.2.1 *We allow the more general case of conditionally positive definite functions for constructing s_g, but we need to restrict to the case of strictly positive kernels for s_h. This enables us to make the eigen-rational interpolant well-defined, that is, such that $s_h(\boldsymbol{x}) \neq 0$, for all $\boldsymbol{x} \in \Omega$.*

Once we provide the function values $s_h(\boldsymbol{x}_i) = h_i$, $i = 1, \ldots, N$, we can construct s_g in the standard way, that is, such that it interpolates $\boldsymbol{g} = (f_1 h_1, \ldots, f_N h_N)^{\intercal}$. Then, obviously, $\hat{s}_f$ interpolates the given function values $\{f_1, \ldots, f_N\}$ at the nodes X.

Since many conditionally positive definite functions $\varphi(r)$ of order m inherit their properties from the m-th derivative of $\varphi(\sqrt{r})$, if the numerator is a conditionally positive definite function of order m, then the associated kernel we choose for the denominator is

$$\frac{d^m}{dt^m} \varphi\left(\sqrt{t}\right)_{t=r^2}.$$

Example 6.2.1 *If φ is the Generalized MultiQuadric, then the associate $\bar{\kappa}$ is the Inverse MultiQuadric kernel. If κ is strictly positive definite, we fix $\bar{\kappa} = \kappa$.*

To show under which conditions the rational interpolant is well-defined, that is, such that $s_h(\boldsymbol{x}) \neq 0$ for all $\boldsymbol{x} \in \Omega$, we need to introduce the *Perron Theorem* Perron (1907).

Definition 6.2.1 A square matrix is called positive when all entries are positive.

Using this definition, the following theorems find application in our setting (Shapiro, 2016).

Theorem 6.2.1 *Every positive square matrix has a positive eigenvalue and the corresponding eigenvector has all positive components.*

The eigenpair, eigenvalue, and eigenvector of this theorem are known as *Perron eigenpair*.

Theorem 6.2.2 *Each positive square matrix possesses exactly one Perron eigenpair. Among all the (real or possibly complex) eigenvalues, the Perron eigenvalue has the largest modulus.*

We can now discuss when the rational interpolant is well-defined. The coefficients $\boldsymbol{b} = (b_1, \dots, b_N)^\mathsf{T}$ are selected so that $\boldsymbol{b}$ is the eigenvector associated with the largest eigenvalue of $\mathsf{K}_{\bar{\kappa}}$, that is, such that it satisfies

$$\max_{\|\tilde{\boldsymbol{b}}\|=1} \tilde{\boldsymbol{b}}^\mathsf{T} \mathsf{K}_{\bar{\kappa}} \tilde{\boldsymbol{b}}, \tag{6.1}$$

where $\mathsf{K}_{\bar{\kappa}}$ is the kernel matrix generated by $\bar{\kappa}$. Of course (6.1) reduces to finding the eigenvector $\boldsymbol{b}$ associated with the largest eigenvalue of $\mathsf{K}_{\bar{\kappa}}$.

Thus, in our case, if the kernel $\bar{\kappa}$ is such that $\bar{\kappa}(\boldsymbol{x}, \cdot) > 0$, for all $\boldsymbol{x} \in \Omega$, by virtue of Theorem 6.2.2, we can conclude that $s_h(\boldsymbol{x}) \neq 0$ for all $\boldsymbol{x} \in \Omega$.

Example 6.2.2 *We have well-defined problems for the Gaussians or Generalized MultiQuadrics. We can also accommodate the compactly supported kernels under certain restrictions. Indeed, one can define* $\boldsymbol{b}$ *as in* (6.1) *subject to the constraint that* b_i *be positive for all* $i = 1, \dots, N$.

Once we compute the vector $\boldsymbol{b}$, we calculate the function values $s_h(\boldsymbol{x}_i) = h_i$, $i = 1, \dots, N$. Then, assuming that κ is a conditionally positive definite radial kernel of order m on $\mathbb{R}^d$, we construct the eigen-rational interpolant $\hat{s}_f$ by solving the usual linear system, where the vector of function values is replaced by $(\boldsymbol{g}, \boldsymbol{0})^\mathsf{T}$, with $\boldsymbol{g} = \boldsymbol{f}\boldsymbol{h}$ and $\boldsymbol{0}$ is a zero vector of length M.

We are now able to construct a data-dependent rational basis. Indeed, let us consider for simplicity a strictly positive definite kernel κ, then the eigen-rational interpolant $\hat{s}_f$ reads as follows:

$$\hat{s}_f = \sum_{j=1}^{N} a_j \frac{\kappa(\boldsymbol{x}, \boldsymbol{x}_j)}{\sum_{i=1}^{N} b_i \kappa(\boldsymbol{x}, \boldsymbol{x}_i)} = \sum_{j=1}^{N} a_j \frac{h_j \kappa(\boldsymbol{x}, \boldsymbol{x}_j)}{\sum_{i=1}^{N} b_i \kappa(\boldsymbol{x}, \boldsymbol{x}_i) \sum_{i=1}^{N} b_i \kappa(\boldsymbol{x}_j, \boldsymbol{x}_i)},$$

where we used $h_j = \sum_{i=1}^{N} b_i \kappa(\boldsymbol{x}_j, \boldsymbol{x}_i)$. Therefore,

$$\hat{s}_f = \sum_{j=1}^{N} \tilde{a}_j \frac{\kappa(\boldsymbol{x}, \boldsymbol{x}_j)}{\sum_{i=1}^{N} b_i \kappa(\boldsymbol{x}, \boldsymbol{x}_i) \sum_{i=1}^{N} b_i \kappa(\boldsymbol{x}_j, \boldsymbol{x}_i)} =: \sum_{j=1}^{N} \tilde{a}_j \kappa_R(\boldsymbol{x}, \boldsymbol{x}_j).$$

Since s_h does not vanish, the function

$$\kappa_R(\boldsymbol{x}, \boldsymbol{y}) = \frac{1}{s_h(\boldsymbol{x})} \frac{1}{s_h(\boldsymbol{y})} \kappa(\boldsymbol{x}, \boldsymbol{y}),$$

is strictly positive definite (see Aronszajn, 1950 and De Marchi et al., 2016). The same argument applies when κ is only conditionally positive definite of order m. In that case, κ_R is conditionally positive definite of order m.

6.3 Stability Analysis

Lebesgue constants are stability indicators that should be studied for a better understanding of the behavior of the error. To this aim, we need the cardinal form for our eigen-rational interpolant.

Proposition 6.3.1 *Suppose κ is a conditionally positive definite kernel of order m in $\mathbb{R}^d$ and $\bar{\kappa}$ is the associated strictly positive definite kernel. Suppose $X = \{\boldsymbol{x}_i, i = 1,\ldots,N\} \subset \Omega$ is $(m-1)$-unisolvent, then there exist functions $\hat{u}_k$, $k = 1,\ldots,N$, such that*

$$\hat{s}_f(\boldsymbol{x}) = \textstyle\sum_{j=1}^{N} f_j \hat{u}_j(\boldsymbol{x}).$$

Furthermore, if $\kappa = \bar{\kappa}$ is strictly positive definite, the $\hat{u}_k$, $k = 1,\ldots,N$, form a partition of unity.

Proof. If κ is a conditionally positive definite kernel with respect to $\mathbb{P}_{m-1}(\mathbb{R}^d)$ and if $X = \{\boldsymbol{x}_i,\ i = 1,\ldots,N\} \subset \Omega$ is $(m-1)$-unisolvent, from Theorem 4.1.1 (extended to conditionally positive definite kernels), we know that there exist functions

$$u_k \in \mathbb{P}_{m-1}(\mathbb{R}^d) + \left\{ \sum_{j=1}^{N} a_j \kappa(\cdot,\boldsymbol{x}_j) \,:\, \sum_{j=1}^{N} m_j p_j(\boldsymbol{x}_j) = 0,\ p \in \mathbb{P}_{m-1}(\mathbb{R}^d) \right\},$$

such that $u_k(\boldsymbol{x}_i) = \delta_{ik}$. If the kernel $\bar{\kappa}$ is strictly positive definite, the same argument holds for functions $\bar{u}_k \in \operatorname{span}\{\bar{\kappa}(\cdot,\boldsymbol{x}_j),\ \ j = 1,\ldots,N\}$. Therefore,

$$s_g(\boldsymbol{x}) = \sum_{k=1}^{N} g(\boldsymbol{x}_k)\, u_k(\boldsymbol{x}), \quad s_h(\boldsymbol{x}) = \sum_{k=1}^{N} h(\boldsymbol{x}_k)\, \bar{u}_k(\boldsymbol{x}), \quad \boldsymbol{x} \in \Omega.$$

Thus, the resulting eigen-rational interpolant in cardinal form is given by

$$\begin{aligned}
\hat{s}_f(\boldsymbol{x}) &= \frac{\sum_{j=1}^{N} g_j u_j(\boldsymbol{x})}{\sum_{k=1}^{N} h_k \bar{u}_k(\boldsymbol{x})} = \frac{\sum_{j=1}^{N} h_j f_j u_j(\boldsymbol{x})}{\sum_{k=1}^{N} h_k \bar{u}_k(\boldsymbol{x})} \\
&= \textstyle\sum_{j=1}^{N} f_j \dfrac{h_j u_j(\boldsymbol{x})}{\sum_{k=1}^{N} h_k \bar{u}_k(\boldsymbol{x})} =: \textstyle\sum_{j=1}^{N} f_j \hat{u}_j(\boldsymbol{x}),
\end{aligned}$$

and, furthermore, $\hat{u}_i(\boldsymbol{x}_i) = \delta_{ik}$, $\boldsymbol{x}_i \in X$. Moreover, if $\kappa = \bar{\kappa}$ is strictly positive definite, $\{\hat{u}_j\}_{j=1}^N$ form a partition of unity, that is, for $\boldsymbol{x} \in \Omega$

$$\sum_{j=1}^N \hat{u}_j(\boldsymbol{x}) = \sum_{j=1}^N h_j \frac{\bar{u}_j(\boldsymbol{x})}{\sum_{k=1}^N h_k \bar{u}_k(\boldsymbol{x})} = \frac{\sum_{j=1}^N h_j \bar{u}_j(\boldsymbol{x})}{\sum_{k=1}^N h_k \bar{u}_k(\boldsymbol{x})} = 1.$$

□

Once we have a cardinal form for the eigen-rational interpolant, we can define the Lebesgue functions $\hat{\lambda}_N$ and constants $\hat{\Lambda}_N$ and provide the following stability estimate:

$$\begin{aligned} |\hat{s}_f(\boldsymbol{x})| = \left| \sum_{i=1}^N f(\boldsymbol{x}_i) \hat{u}_i(\boldsymbol{x}) \right| &\leq \left(\sum_{i=1}^N |\hat{u}_i(\boldsymbol{x})| \right) \|f\|_\infty =: \hat{\lambda}_N(\boldsymbol{x}) \|f\|_\infty \\ &\leq \hat{\Lambda}_N \|f\|_\infty, \end{aligned}$$

where $\hat{\Lambda}_N = \max_{\boldsymbol{x} \in \Omega} \hat{\lambda}_N(\boldsymbol{x})$.

Remark 6.3.1 *As in De Marchi et al. (2016), the cardinal functions of the eigen-rational interpolant are essentially a rescaled form of the ones of the standard RBF approximant. The same holds for the Lebesgue functions, and so we expect that their behavior is similar to that of the standard interpolants.*

6.4 Error Analysis

To formulate error bounds, we need to think of $\boldsymbol{g}$ and $\boldsymbol{h}$ as function values obtained by sampling two functions $g = fh$ and h belonging to $\mathcal{N}_\kappa(\Omega)$ and $\mathcal{N}_{\bar{\kappa}}(\Omega)$, respectively.

Proposition 6.4.1 bounds the error from above for the eigen-rational interpolant in terms of the power function.

Proposition 6.4.1 *Let $\Omega \subseteq \mathbb{R}^d$ be open. Suppose $\kappa \in \mathscr{C}(\Omega \times \Omega)$ be a conditionally positive definite kernel of order m and $\bar{\kappa}$ the associated strictly positive definite kernel. Assume that $X = \{\boldsymbol{x}_i, i = 1, \ldots, N\} \subset \Omega$ is an $(m-1)$-unisolvent set. Then, for $\boldsymbol{x} \in \Omega$, the error between f and its eigen-rational interpolant can be bounded by*

$$|f(\boldsymbol{x}) - \hat{s}_f(\boldsymbol{x})| \leq \frac{1}{|s_h(\boldsymbol{x})|} \left(P_{X,\bar{\kappa}}(\boldsymbol{x}) \|h\|_{\mathcal{N}_{\bar{\kappa}}(\Omega)} |f(\boldsymbol{x})| + P_{X,\kappa}(\boldsymbol{x}) \|g\|_{\mathcal{N}_\kappa(\Omega)} \right).$$

Proof. Let us consider $\boldsymbol{x} \in \Omega$, we have that

$$|f(\boldsymbol{x}) - \hat{s}_f(\boldsymbol{x})| = \left| \frac{s_h(\boldsymbol{x}) f(\boldsymbol{x}) - s_g(\boldsymbol{x})}{s_h(\boldsymbol{x})} \right|,$$

$$= \left| \frac{(s_h(\boldsymbol{x}) f(\boldsymbol{x}) - h(\boldsymbol{x}) f(\boldsymbol{x})) + (g(\boldsymbol{x}) - s_g(\boldsymbol{x}))}{s_h(\boldsymbol{x})} \right|.$$

Then by applying Theorem 4.1.2, we obtain

$$|f(\boldsymbol{x}) - \hat{s}_f(\boldsymbol{x})| \leq \frac{1}{|s_h(\boldsymbol{x})|} \left(P_{X,\bar{\kappa}}(\boldsymbol{x}) \|h\|_{\mathcal{N}_{\bar{\kappa}}(\Omega)} |f(\boldsymbol{x})| + P_{X,\kappa}(\boldsymbol{x}) \|g\|_{\mathcal{N}_{\kappa}(\Omega)} \right),$$

as claimed. □

The statement of Proposition 6.4.1 enables us to bound the error in terms of the power functions defined on X associated to κ and $\bar{\kappa}$. However, to test the method's convergence, we need to investigate how the errors depend on the fill distance (see Definition 4.1.3).

Proposition 6.4.2 *Suppose $\Omega \subseteq \mathbb{R}^d$ is open and bounded and satisfies an interior cone condition. Further, let $\kappa \in \mathscr{C}^{2k}(\Omega \times \Omega)$ be symmetric and conditionally positive definite of order m and $\bar{\kappa}$ the associated strictly positive definite kernel. Assume that $X = \{\boldsymbol{x}_i, i = 1, \ldots, N\} \subset \Omega$ is an $(m-1)$-unisolvent set. Then there exist positive constants h_0 and $\tilde{C}$ independent of $\boldsymbol{x}$, f, and κ, such that*

$$|f(\boldsymbol{x}) - \hat{s}_f(\boldsymbol{x})| \leq \frac{\tilde{C} h_{X,\Omega}^k \sqrt{\tilde{C}_{\kappa^*,\boldsymbol{x}}}}{|s_h(\boldsymbol{x})|} \left(\|h\|_{\mathcal{N}_{\bar{\kappa}}(\Omega)} \|f(\boldsymbol{x})\| + \|g\|_{\mathcal{N}_{\kappa}(\Omega)} \right)$$

provided $h_{X,\Omega} \leq h_0$, with

$$\tilde{C}_{\kappa^*,\boldsymbol{x}} = \max\left(C_{\varphi,\boldsymbol{x}}, C_{\bar{\kappa},\boldsymbol{x}} \right),$$

and where $C_{\kappa,\boldsymbol{x}}$ and $C_{\bar{\kappa},\boldsymbol{x}}$ are computed as in Theorem 4.1.3.

Remark 6.4.1 *Theoretically, the eigen-rational interpolant and the standard one show similar convergence rates. However, the eigen-rational interpolant takes numerical advantage of rescaling, resulting in more accurate computations.*

6.5 Numerical Experiments

In the numerical tests, at first we focus on the Lebesgue functions and then on the convergence rates. In the first tests, we compare the Lebesgue functions λ_N of the standard RBF interpolant with the corresponding ones of the eigen-rational approximant, that is, $\hat{\lambda}_N$. We calculate the Lebesgue functions for $N = 10$ and $d = 1$ for different kinds of nodes (Halton and Chebyshev); refer to Figure 6.1. Note that the behavior of the Lebesgue functions is comparable.

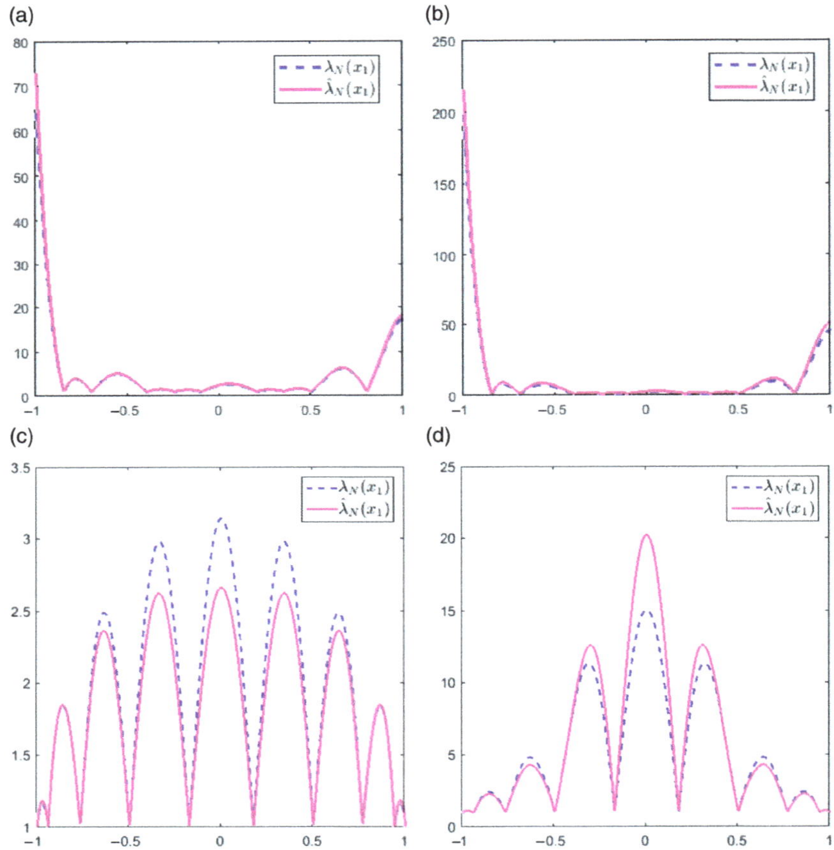

Figure 6.1 Lebesgue functions λ_N and $\hat{\lambda}_N$ for standard and eigen-rational RBF interpolants, respectively. They are computed for different kinds of datasets on $\Omega = [-1, 1]$ with $N = 10$. Halton points computed via the Generalized MultiQuadric kernel with $\varepsilon = 0.5$ (a) and $\varepsilon = 3$ (b). Chebyshev points computed via the Gaussian kernel with $\varepsilon = 0.5$ (c) and $\varepsilon = 3$ (d).

To numerically study the errors, we evaluate the eigen-rational interpolants on $v = 40^d$ equally spaced points $\{z_i, i = 1, \dots, M\}$. Let f be a function on Ω and let ι be an interpolant of f. Common error indicators are the maximum absolute error (MAE) and the root mean square error (RMSE)

$$\text{MAE} = \max_{1 \le k \le M} |f(z_k) - \iota(z_k)|, \; \text{RMSE} = \sqrt{\frac{1}{M} \sum_{k=1}^{M} (f(z_k) - \iota(z_k))^2}. \quad (6.2)$$

For comparing the convergence rates of eigen-rational and standard interpolants, we need to study the error by varying the number of nodes N and evaluate the *empirical convergence rates*

$$\Gamma_k = \frac{\log\left(\text{RMSE}_{k-1}/\text{RMSE}_k\right)}{\log\left(h_{X^{k-1},\Omega}/h_{X^k,\Omega}\right)}, \quad k = 2, 3, \dots,$$

where the index k denotes the k-th numerical experiment, and $h_{X^k,\Omega}$ is the fill distance of the k-th computational mesh. We consider several nested sets of Halton and grid points on $\Omega = [0,1]^2$ and the following test function

$$f(x_1, x_2) = \text{sinc}(x_1)\text{sinc}(x_2).$$

The results of these experiments are reported in Table 6.1. We recover spectral convergence rates for $\mathscr{C}^\infty$ RBFs, while for functions with finite regularities, such as the considered Buhmann $\mathscr{C}^3$ RBF given by $\varphi(r) = 112/45r^{9/2} + 16/3r^{7/2} - 7r^4 - 14/15r^2 + 1/9$, the convergence rates are limited by the order of smoothness of the kernel. Furthermore, note that despite such rates being comparable, the eigen-rational interpolant is found to be more accurate, especially for the Gaussian function. For such a particular RBF, it is also possible to see faster convergence rates.

Table 6.1 RMSEs and convergence rates for standard and eigen-rational interpolants, respectively, computed with f on five sets of grid data on $\Omega = [0,1]^2$

	Gaussian ($\varepsilon = 3$)				Buhmann $\mathscr{C}^3$			
	Classical RBF		Rational RBFs		Rational RBFs		Classical RBFs	
N	RMSE	Γ	RMSE	Γ	RMSE	Γ	RMSE	Γ
25	1.76E − 02	-	1.69E − 03	-	1.04E − 02	-	2.04E − 03	-
49	3.29E − 03	3.64	2.15E − 04	4.48	3.62E − 03	2.29	4.50E − 04	3.29
81	4.95E − 04	8.10	1.41E − 05	11.6	1.61E − 03	3.45	1.73E − 04	4.09
289	1.12E − 07	12.1	1.19E − 11	20.1	2.01E − 04	2.99	1.91E − 05	3.17
1089	2.73E − 08	2.03	3.70E − 15	11.5	1.33E − 04	3.91	1.17E − 06	4.02

7
Partition of Unity

The PU method addresses key challenges of computational complexity and numerical instability in global kernel-based approximation techniques. Instead of constructing a single global model over the entire domain, the PU approach decomposes the domain into overlapping subdomains, each covering a local region. On each subdomain, a local interpolant is built independently from nearby data points, significantly reducing the size of the local problems and improving numerical conditioning. In this chapter, we present the PU method and its main features.

7.1 Local Techniques

Local techniques, such as the PU scheme, have been used for interpolation since around 1960 (McLain, 1976; Shepard, 1968). Later, they have been coupled with RBFs (Fasshauer, 2007; Wendland, 2002). Moreover, the PU method for solving PDEs, first introduced in the mid-1990s in Babuška and Melenk (1997) and Melenk and Babuška (1996), is now a popular technique (Heryudono and Raessi, 2023; Perracchione, 2017; Safdari-Vaighani et al., 2015; Shcherbakov and Larsson, 2016).

These local interpolants are then combined into a global approximation through a set of smoothly varying weight functions that form a PU. This weighted blending ensures that the final approximation is continuous and coherent across subdomain boundaries. Theoretical analysis demonstrates that PU-based approximants retain the same local approximation order as the underlying local schemes, meaning that no accuracy is lost in the transition from local to global representation. Corresponding error bounds confirm that the global error behaves consistently with local approximation errors.

From a computational perspective, the PU framework, the topic of the present chapter, benefits from efficient data structures, such as integer-based spatial partitioning, that enable fast point location, neighbor searches, and a scalable implementation. These properties make the method particularly suitable for large datasets. The effectiveness of the PU approach is demonstrated through applications such as implicit surface reconstruction from three-dimensional point clouds, where accuracy, robustness, and efficiency are critical.

7.2 The PU Framework

The PU method enables us to handle large datasets. Indeed:

1. The basic idea of the PU method is to start with a partition of the open and bounded domain Ω into $p \in \mathbb{N}$ subdomains or patches Ω_j, such that $\Omega \subseteq \cup_{k=1}^{p} \Omega_k$, with some mild overlap among them.
2. The subdomains must form a covering of the domain and, moreover, the overlap must be sufficient so that each interior point $\boldsymbol{x} \in \Omega$ is located in the interior of at least one patch Ω_k.

Example 7.2.1 *The overlap condition is illustrated in Figure 7.1(a). Specifically, a two-dimensional view of a PU structure, covering a set of scattered data in the unit square and satisfying the properties mentioned above, is shown in the left frame. In this example, we consider circular patches, but other shapes can be used as well. In Figure 7.1(b), we plot a PU structure that does not satisfy the overlap condition.*

According to Wendland (2002), some additional assumptions are required on the regularity of the covering $\{\Omega_k\}_{k=1}^{p}$.

Definition 7.2.1 Suppose that $\Omega \subseteq \mathbb{R}^d$ is bounded and $X = \{\boldsymbol{x}_k,\ k = 1,\dots,N\} \subseteq \Omega$ is given. An open and bounded covering $\{\Omega_k\}_{k=1}^{p}$, $p \in \mathbb{N}$, is called regular for (Ω, X) if the following properties are satisfied:

(i) For each $\boldsymbol{x} \in \Omega$, the number of subdomains Ω_k, with $\boldsymbol{x} \in \Omega_k$, is bounded by a global constant C.
(ii) Each subdomain Ω_k satisfies an interior cone condition.
(iii) The local fill distances $h_{X_{N_k},\Omega_k}$ are uniformly bounded by the global fill distance $h_{X,\Omega}$, where $X_{N_k} = X \cap \Omega_k$.

Associated with these subdomains, a PU, that is, a family of compactly

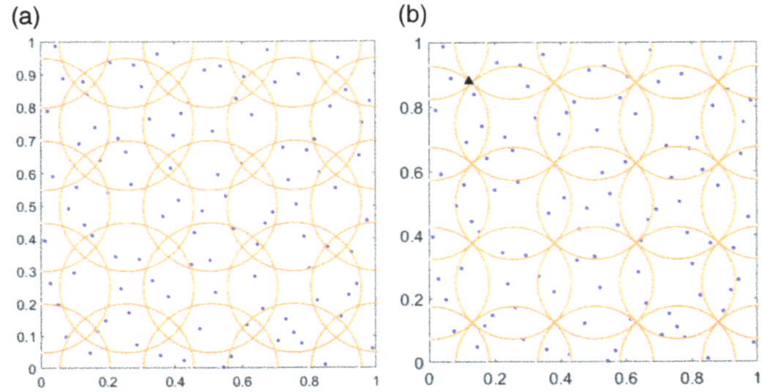

Figure 7.1 An illustrative example of PU subdomains covering the domain $\Omega = [0,1]^2$ and satisfying the overlap condition (a). In (b), the covering does not satisfy the overlap condition; a critical point where no overlap occurs is marked with a triangle. The dots represent scattered data, and the circles identify the PU subdomains.

supported, nonnegative, continuous functions ω_k, with $\text{supp}(\omega_k) \subseteq \Omega_k$ and such that

$$\sum_{k=1}^{p} \omega_k(\boldsymbol{x}) = 1, \quad \boldsymbol{x} \in \Omega$$

is considered. In addition, according to Wendland (2002), we suppose that $\{\omega_k\}_{k=1}^p$ is a α-stable PU, that is, for every $\boldsymbol{\beta} \in \mathbb{N}^d$, with $|\boldsymbol{\beta}| \leq \alpha$, there exists a constant $C_{\boldsymbol{\beta}} > 0$ such that

$$\left\| D^{\boldsymbol{\beta}} \omega_k \right\|_{\mathscr{L}^\infty(\Omega_k)} \leq \frac{C_{\boldsymbol{\beta}}}{\left(\sup_{\boldsymbol{x},\boldsymbol{y} \in \Omega_k} \|\boldsymbol{x} - \boldsymbol{y}\|\right)^{|\boldsymbol{\beta}|}}, \quad k = 1, \ldots, p.$$

Among several weight functions, a possible choice is given by the *Shepard's weights* Shepard (1968), that is

$$\omega_k(\boldsymbol{x}) = \frac{\bar{\omega}_k(\boldsymbol{x})}{\sum_{j=1}^{p} \bar{\omega}_j(\boldsymbol{x})}, \quad k = 1, \ldots, p, \tag{7.1}$$

where $\bar{\omega}_j$ are compactly supported functions, with support on Ω_j. Moreover, such a family $\{\omega_k\}_{k=1}^p$ forms a PU.

Once we choose the PU $\{\omega_k\}_{k=1}^p$, the global interpolant is formed by the weighted sum of p local approximants s_{f_k}, that is,

$$\mathcal{I}(\boldsymbol{x}) = \sum_{k=1}^{p} s_{f_k}(\boldsymbol{x})\,\omega_k(\boldsymbol{x}), \quad \boldsymbol{x} \in \Omega. \tag{7.2}$$

Remark 7.2.1 *We point out the following:*

1. *The assumptions in Definition 7.2.1 lead to the requirement that the number of subdomains is proportional to the number of data.*
2. *The first property ensures that the sum in* (7.2) *is actually a sum over at most C summands. The fact that C is independent of N, unlike p, is essential to avoid loss of convergence orders.*
3. *For an efficient evaluation of the global interpolant, it is crucial that only a constant number of local approximants is evaluated. This means that it should be possible to locate those C indices in constant time.*

In particular, here s_{f_k} denotes an RBF interpolant defined on a subdomain Ω_k. For instance, supposing to have a positive definite function, the k-th local interpolant assumes the form

$$s_{f_k}(\boldsymbol{x}) = \sum_{j=1}^{N_k} c_j^k \varphi(\|\boldsymbol{x} - \boldsymbol{x}_j^k\|), \tag{7.3}$$

where N_k indicates the number of points in Ω_k and $\boldsymbol{x}_j^k \in X_{N_k}$, with $j = 1, \ldots, N_k$. The coefficients $\{c_j^k\}_{j=1}^{N_k}$ in (7.3) are determined by enforcing the N_k local interpolation conditions

$$s_{f_k}(\boldsymbol{x}_j^k) = f_j^k, \quad j = 1, \ldots, N_k.$$

Thus, in the case of positive definite functions, the problem of finding the PU interpolant (7.2) reduces to solving p linear systems of the form

$$\mathsf{K}_j \boldsymbol{c}_j = \boldsymbol{f}_j,$$

where $\boldsymbol{c}_j = (c_1^j, \ldots, c_{N_j}^j)^\intercal$, $\boldsymbol{f}_j = (f_1^j, \ldots, f_{N_j}^j)^\intercal$, and $\mathsf{K}_j \in \mathbb{R}^{N_j \times N_j}$ is

$$\mathsf{K}_j = \begin{pmatrix} \varphi(\|\boldsymbol{x}_1^j - \boldsymbol{x}_1^j\|) & \cdots & \varphi(\|\boldsymbol{x}_1^j - \boldsymbol{x}_{N_j}^j\|) \\ \vdots & \ddots & \vdots \\ \varphi(\|\boldsymbol{x}_{N_j}^j - \boldsymbol{x}_1^j\|) & \cdots & \varphi(\|\boldsymbol{x}_{N_j}^j - \boldsymbol{x}_{N_j}^j\|) \end{pmatrix}.$$

Since the functions ω_k, $k = 1, \ldots, p$ form a PU, if the local fits s_{f_k} satisfy the interpolation conditions, then the global PU approximant inherits the interpolation property (Fasshauer, 2007; Wendland, 2002).

7.2.1 Error Bounds for Radial Basis Function Partition of Unity Interpolants

To formulate error bounds, we need to define the space $\mathscr{C}_\nu^\alpha(\mathbb{R}^d)$ of all functions $f \in \mathscr{C}^\alpha$ whose derivatives of order $|\boldsymbol{\beta}| = \alpha$ satisfy $D^{\boldsymbol{\beta}} f(\boldsymbol{x}) = \mathscr{O}(||\boldsymbol{x}||^\nu)$ for $||\boldsymbol{x}|| \longrightarrow 0$. We are now able to give a convergence result (cf. Wendland, 2002).

Theorem 7.2.1 *Let $\Omega \subseteq \mathbb{R}^d$ be open and bounded and suppose that $X = \{\boldsymbol{x}_i, i = 1,\dots,N\} \subseteq \Omega$. Let $\phi \in \mathscr{C}_\nu^\alpha(\mathbb{R}^d)$ be a positive definite function. Let $\{\Omega_k\}_{k=1}^P$ be a regular covering for (Ω, X) and let $\{\omega_k\}_{k=1}^P$ be α-stable for $\{\Omega_k\}_{k=1}^P$. Then the error between $f \in \mathscr{N}_\kappa(\Omega)$ and its PU interpolant* (7.2) *can be bounded by*

$$|D^{\boldsymbol{\beta}} f(\boldsymbol{x}) - D^{\boldsymbol{\beta}} \mathscr{I}(\boldsymbol{x})| \leq C' h_{X,\Omega}^{(\alpha+\nu)/2-|\boldsymbol{\beta}|} ||f||_{\mathscr{N}_\kappa(\Omega)},$$

for all $\boldsymbol{x} \in \Omega$ and all $|\boldsymbol{\beta}| \leq \alpha/2$.

If we compare the result reported in Theorem 7.2.1 with the global error estimates shown in Chapters 4 and 5, we can see that the PU interpolant preserves the local approximation order for the global fit. Hence, we can efficiently compute large RBF interpolants by solving smaller interpolation problems and then combining them with the PU $\{\omega_k\}_{k=1}^P$.

7.2.2 Partitioning Data Structures

In the PU setting, an important issue is the *organization of points* across the subdomains. In literature, techniques such as *kd-trees* have already been constructed. They

- allow data partitioning in a k-dimensional space, and related searching procedures have already been designed (Arya et al., 1998; Berg et al., 1997) and
- work with high dimensions, but they are not specifically implemented for the PU method.

Alternatively, schemes for multivariate interpolation make use of the so-called *Integer-based Partitioning Structure (I-PS)* and a related search procedure (Cavoretto et al., 2016). Here are some facts about the partitioning structure:

- It strictly depends on the size of the PU subdomains and allows to deal with a truly large number of data with a relatively low computational complexity.
- It follows from the results shown in Allasia et al. (2011) and Cavoretto and De Rossi (2014, 2012, 2010), where efficient searching procedures based on the partition of the underlying domains in strips or crossed strips are considered.

7.3 Modeling 3D Objects via PU Interpolation: The Implicit Approach

A common problem in computer-aided design and computer graphics is the reconstruction of surfaces from point cloud data, that is, a set of unorganized 3D points. Such applications arise in computer graphics, modeling complicated 3D objects or in medical imaging (see, e.g., Chen and Lai, 2007; Cuomo et al., 2013; Morse et al., 2001; and Zhu and Wathen, 2015).

An approach to obtain a surface that fits the given 3D point cloud data is based on the use of implicit surfaces defined in terms of some meshfree approximation methods, such as RBF interpolant Fasshauer (2007). More details can also be found in Carr et al. (2001, 1997); Drake et al. (2022); Ohtake et al. (2006); Turk and O'Brien (2002); and Yao et al. (2015).

Given a point cloud data set $X = \{\boldsymbol{x}_i \in \mathbb{R}^3,\ i = 1,\dots,N\}$, belonging to an unknown 2D manifold M, namely a surface in $\mathbb{R}^3$, we seek another surface M^* that approximates M.

For the implicit approach, we think of M as the surface of all points $\boldsymbol{x} \in \mathbb{R}^3$ satisfying the implicit equation

$$f(\boldsymbol{x}) = 0,$$

for some function f.

The key point to finding the interpolant of the trivariate function f from the given data points is to use additional significant interpolation conditions, that is, to add an extra set of off-surface points. Once we define the augmented data set, we can then compute a 3D interpolant $\mathcal{I}$, via the PU method, to the total set of points (Fasshauer, 2007; Wendland, 2002).

Thus, the reconstruction of the surface leads to a method consisting of three steps:

1. Generate the extra off-surface points.
2. Find the interpolant of the augmented data set.
3. Render the iso-surface of the fit.

Suppose that for each point $\boldsymbol{x}_i$, the oriented normal $\boldsymbol{n}_i \in \mathbb{R}^3$ is given. We construct the extra off-surface points by taking a small step away along the surface normals, that is, we obtain for each data point $\boldsymbol{x}_i$ two additional off-surface points:

$$\boldsymbol{x}_{N+i} = \boldsymbol{x}_i + \Delta \boldsymbol{n}_i, \quad \boldsymbol{x}_{2N+i} = \boldsymbol{x}_i - \Delta \boldsymbol{n}_i,$$

Δ being the stepsize.

The union of the sets $X_\Delta^+ = \{\boldsymbol{x}_{N+1},\dots,\boldsymbol{x}_{2N}\}$, $X_\Delta^- = \{\boldsymbol{x}_{2N+1},\dots,\boldsymbol{x}_{3N}\}$, and X, namely X_{3N}, gives the overall set of points on which the interpolation conditions are assigned.

In what follows, for a given $\Delta > 0 \in \mathbb{R}^+$, we fix $\Delta_i = \Delta$, $i = 1,\ldots,N$. More precisely, to possibly avoid the ill-conditioning effect, the extended dataset can be constructed with a stepsize Δ proportional to the separation distance q_X of the original dataset. In particular, we take $\Delta = \xi q_X$, with $\xi = 1/3$. This choice is based on the investigation carried out by Cuomo et al. (2017). Indeed, the authors prove that for a suitable choice of the stepsizes $\Delta_i > 0 \in \mathbb{R}^+, i = 1,\ldots,N$ and for $0 < \xi \leq 1/3$ we have $q_{X_N} = \xi q_X$. Therefore, since ill-conditioning increases with decreasing q_X, it is not recommended to take small values for ξ. On the other hand, for $\xi > 1/3$ the added points might be close to each other, and this might lead to the phenomenon of self-intersecting surfaces. Therefore, fixing $\xi = 1/3$ seems to provide a good compromise between accuracy and stability.

Note that if we have zero normals in the given normal data set, we must exclude such points. Finally, we construct the augmented set of function values. It is defined as the union of the following sets:

$$
\begin{aligned}
F_N &= \{f_i \ : \ f(\boldsymbol{x}_i) = a, \ i = 1,\ldots,N\},\\
F_\Delta^+ &= \{f_i \ : \ f(\boldsymbol{x}_i) = b, \ i = N+1,\ldots,2N\},\\
F_\Delta^- &= \{f_i \ : \ f(\boldsymbol{x}_i) = c, \ i = 2N+1,\ldots,3N\},
\end{aligned}
$$

where the values of a, b, and $c \in \mathbb{R}$ are arbitrarily and usually set as 0, 1, and -1, respectively. Now that we have created the data set, we can compute an interpolant whose a-contour (iso-surface) interpolates the given point cloud. Note that this approach can be extended to rational RBFs, as discussed in Perracchione (2018). If the normals are not explicitly provided, see, e.g., Hoppe et al. (1992) and Hoppe (1994) for estimating them.

Example 7.3.1 *Let us consider the following data set*

$$
\boldsymbol{x}_i = \left([2+\sin(t_i)]\cos(t_i), [2+\cos(t_i)]\sin(t_i)\right), \quad i = 1,\ldots,N,
$$

where t_i is a Halton sequence .

Let us fix $N = 75$. In Figure 7.2, we show the points and the normals to the curve.

Next, to obtain the set of off-surface points, we add the function values. Specifically, we assign the value 0 to each original data point and the value 1 or -1 to outside or inside points (obtained by marching a small distance Δ along the normals). Then, we proceed as for standard interpolation, and we recover the curve plotted in Figure 7.3. It is the zero contour of the interpolant surface.

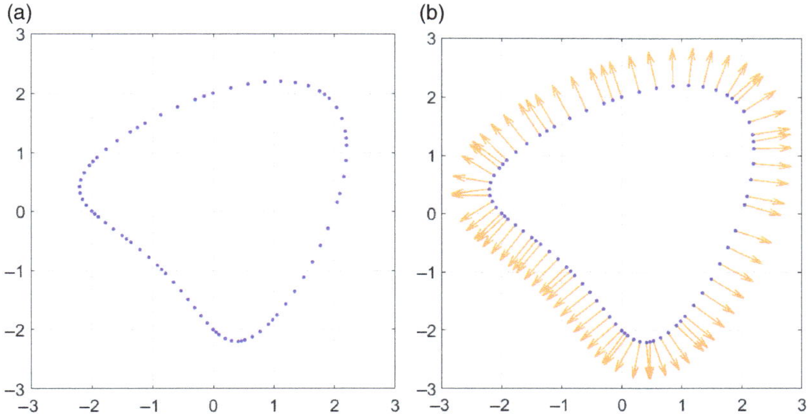

Figure 7.2 (a) The point cloud data set. (b) Consistent set of normals.

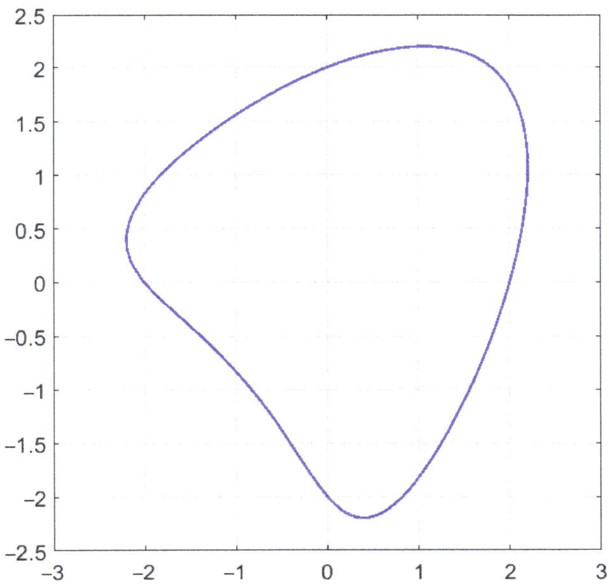

Figure 7.3 The interpolant curve, that is, the zero contour of the interpolant surface.

8
Efficient Reduced-Basis Approximation

In the previous Chapter 7, we presented the PU approach as a suitable strategy for leveraging a potentially large dataset of nodes and simultaneously overcoming computational and instability issues. With a different perspective, a remarkable amount of recent and older research has been devoted to the problem of selecting a suitable subset of nodes from the larger set at hand. Indeed, it is possible to construct *lighter* kernel-based surrogate models s_{X_n} that can still retain an accuracy comparable to that of the interpolant built at the whole dataset s_{X_N}, or behave even better, with $X_n \subset X_N$ and $n \ll N$. In this context, many efforts have been devoted to the study of iterative methods, in which the final approximation design is extracted from the original dataset through multiple steps.

In this direction, there are two possible approaches: starting from the full dataset and iteratively removing nodes, or starting from a small or empty set and iteratively adding nodes from the dataset, according to a selection criterion. We will focus on the first setting. The latter captures the basic idea of *greedy* kernel methods, briefly reviewed in the following section.

8.1 A Note on Greedy Algorithms

Greedy procedures have been thoroughly analyzed in the literature (Albrecht and Iske, 2025; Wenzel et al., 2025; Su and Ling, 2025), both from a theoretical and a computational perspective (De Marchi et al., 2005; Dutta et al., 2021; Schaback, 2000; Wenzel et al., 2023; Wirtz et al., 2014). In particular, the Vectorial Kernel Orthogonal Greedy Algorithm (VKOGA) (Santin and Haasdonk, 2021; Wirtz and Haasdonk, 2013) implements several greedy algorithms efficiently. We outline the main steps of VKOGA next.

Start with an empty set $X_0 = \{\}$. Then for $n \geq 1$, at the n-th step the set X_n is defined as $X_n = X_{n-1} \cup \{\boldsymbol{x}_n\}$ and $\boldsymbol{x}_n$ is so that

$$\boldsymbol{x}_n := \text{argmax}_{\boldsymbol{x} \in X_N \setminus X_{n-1}} \eta^{(n)}(\boldsymbol{x}),$$

using some error indicator $\eta^{(n)} : \Omega \longrightarrow \mathbb{R}$.
In the kernel literature, the following criteria are frequently used (De Marchi et al., 2005; Schaback, 2000), and they make use of the residual $r_{X_n} \equiv f - s_{X_n}$ or the power function (see Theorem 4.1.2) or both:

1. P-greedy: $\eta_P^{(n)}(\boldsymbol{x}) = P_{X_n,\kappa}(\boldsymbol{x})$,
2. f-greedy: $\eta_f^{(n)}(\boldsymbol{x}) = |r_{X_n}(\boldsymbol{x})|$,
3. f/P-greedy: $\eta_{f/P}^{(n)}(\boldsymbol{x}) = |r_{X_n}(\boldsymbol{x})|/P_{X_n,\kappa}(\boldsymbol{x})$.

The convergence rates for the P-greedy algorithm were analyzed in Santin and Haasdonk (2017) and Wenzel et al. (2021, 2024b) and, recently, the previous criteria were unified and analyzed within the scale of so-called *β-greedy algorithms* (Wenzel et al., 2023).

Remark 8.1.1 *Target data-dependent algorithms, such as f-greedy, achieve faster convergence rates by exploiting information from the underlying function f.*

8.2 Nodes Removal Made Efficient

One reason for the success of greedy algorithms in the literature is their efficient computation. Indeed, if $n \ll N$, searching for the node to be added in the next step is not a big issue. On the other hand, if one starts from the full dataset and wants to remove nodes iteratively, a straightforward implementation leads to the construction of many *heavy* models. In what follows, we present schemes from Marchetti and Perracchione (2022a) for reduced basis models based on the minimization of both the residual and the power function, which are made efficient by the usage of the ERA (see Section 5.2). First, we present the steps of the nodes removal scheme, which shows similarities with the *knot removal* algorithm discussed in (Fasshauer, 2007, section 21). We will allow removing more than one node at each step of the algorithm. Then, we focus on its fast implementation. The resulting approach is called the efficient reduced basis algorithm (ERBA).

Let X_N be the initial set of nodes, let $X_{t-1} = \{\boldsymbol{x}_i,\ i = 1, \dots, n_{t-1}\} \subset \Omega$ be the reduced set at the $(t-1)$-th step of the algorithm ($X_0 = X$), and let $\tau \in \mathbb{R}_{>0}$ be a fixed tolerance. Choosing $\rho \in \mathbb{N}$, $\rho < n$, and letting $\ell = \lfloor n_{t-1}/\rho \rfloor$, the t-th step of the iterative scheme, $t \geq 1$, is as follows.

1. Partition X_{t-1} into ℓ test sets $X_{t-1}^1, \dots, X_{t-1}^\ell$, $|X_{t-1}^j| = \rho^j$ with $\rho^j \in \{\rho, \dots, 2\rho - 1\}$, $\rho^j \le \max(2\rho - 1, n_{t-1})$. Moreover, let $\overline{X}_{t-1}^j := X_{t-1} \setminus X_{t-1}^j$ be the training (interpolation) sets, $j = 1, \dots, \ell$.
2. For each X_{t-1}^j, $j = 1, \dots, \ell$, compute:

2a. for the residual-based scheme:

$$w^j = \frac{1}{\sqrt{\rho^j}} \|\boldsymbol{f}^j - \boldsymbol{s}^j\|_2,$$

where $\boldsymbol{f}^j := \boldsymbol{f}_{|_{X_{t-1}^j}}$ and $\boldsymbol{s}^j := s_{\overline{X}_{t-1}^j}(X_{t-1}^j)$ is the evaluation vector on X_{t-1}^j of the interpolant $s_{\overline{X}_{t-1}^j}$;

2b. for the power-based scheme:

$$w^j = \frac{1}{\sqrt{\rho^j}} \|\boldsymbol{P}^j\|_2,$$

where $\boldsymbol{P}^j := P_{\overline{X}_{t-1}^j, \kappa}(X_{t-1}^j)$ is the evaluation vector at X_{t-1}^j of the power function $P_{\overline{X}_{t-1}^j, \kappa}$;

3. Choose

$$j^\star = \operatorname{argmin}_{j \in \{1, \dots, \ell\}} w^j.$$

4. Then:

4a. if $w^{j^\star} \le \tau$, define $X_t = X_{t-1} \setminus X_{t-1}^{j^\star}$ and proceed iteratively with the t-th step;

4b. if $w^{j^\star} > \tau$, $X_t = X_{t-1}$ is the final interpolation set obtained by the procedure.

Remark 8.2.1 *In the presented algorithm, the data set is randomly partitioned at each iteration. Since $\rho \ll n$, such randomness does not influence the algorithm in selecting meaningful designs of interpolation node sets, as also confirmed by the numerical experiments carried out in Marchetti and Perracchione (2022b).*

As for convergence, the residual-based rule has been studied in many contexts (see, e.g., Lyche, 1992). As far as the convergence of the proposed power function-based scheme is concerned, it is a consequence of (Santin and Haasdonk, 2017, Lemma 2.3), that is,

$$\|P_{X_{t-1},\kappa}\|_{\mathscr{L}_\infty(\Omega)} \le \|P_{\kappa, X_t}\|_{\mathscr{L}_\infty(\Omega)},$$

being $X_t \subset X_{t-1}$ nested sets. We can assume without loss of generality that ρ divides n_t at each step of the algorithm, and thus $\rho_j \equiv \rho$ holds true. Therefore, at each step, a classical implementation of this method requires solving n_t

different $(n_t - \rho) \times (n_t - \rho)$ linear systems, in the residual-based case, or the inversion of $(n_t - \rho) \times (n_t - \rho)$ matrices in the power-based setting. Note that the computational bottleneck is in Step 2 of the algorithm.

In the residual-based scheme, Step 2a can be sped-up by directly using the ERA from Section 5.2 to compute the residual. In this way, we recall that we need to invert an $n_t \times n_t$ matrix once, and then to solve ℓ different small $\rho \times \rho$ linear systems, for a computational cost of $\mathscr{O}(n_t^3)$.

For the power-based case, inspired by the ERA, in (Marchetti and Perracchione, 2022a, Theorem 1), a solution for a faster computation of the vector of the power function has been proposed. With that solution, the computational cost at each step is reduced to $\mathscr{O}(n_t^3)$ as in the residual-based scheme.

8.3 Testing ERBA

Here, we provide a brief overview of the efficiency of the ERBA approach by presenting some numerical tests. In all the experiments, we take $\Omega = [-1,1]^2$ and the $\mathscr{C}^0$ Matérn strictly positive definite kernel. As interpolation data set $X \subset \Omega$, we consider an $n \times n$ grid, with $n = 25$. In addition, the results are evaluated on a 60×60 evaluation grid, Ξ. We compare ERBA to RBA, which is the reduced-basis algorithm where Step 2 is not optimized by using ERA.

We test two target functions. The regular $f : \Omega \longrightarrow \mathbb{R}$ defined as

$$f(\boldsymbol{x}) = \frac{1}{1 + (x_1 - 0.5)^2 + (x_2 + 0.2)^2}, \qquad \boldsymbol{x} = (x_1, x_2),$$

and the discontinuous $g : \Omega \longrightarrow \mathbb{R}$ defined as

$$g(\boldsymbol{x}) = \begin{cases} x_1 + x_2 - 3 & \text{if } x_1 > 0, \\ x_1 + x_2 - 2 & \text{if } x_1 \leq 0, \end{cases} \qquad \boldsymbol{x} = (x_1, x_2).$$

In Figure 8.1, we display f and the error calculated by the interpolant constructed at the whole set X. In Table 8.1, as well as in Figure 8.2, we present the results related to both residual- and power-based ERBA, in terms of accuracy and computational time, in interpolating f. The results related to g are shown analogously in Figures 8.3 and 8.4 and in Table 8.2.

We note that the usage of the power-based scheme leads to quasi-uniform interpolation designs X_t. This is a reasonable effect that is also present in greedy approaches. On the other hand, when using information on the underlying function with the residual-based algorithm, the final distribution of the nodes is heavily influenced by the behavior of the target function.

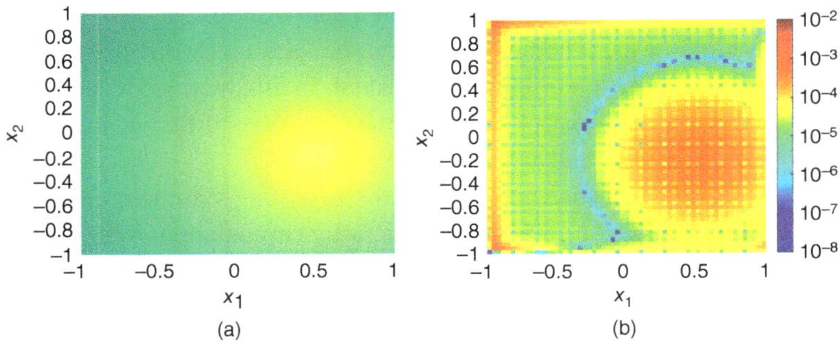

Figure 8.1 (a) The function f. (b) The pointwise relative error between f and the interpolant constructed via the full data set X. The associated RMSE computed on Ξ is $9.69\text{E}-05$.

Table 8.1 Results for the residual-based and power-based ERBA schemes with f. CPU times are in seconds

$\lvert X_t \rvert$	RMSE	ERBA CPU times	RBA CPU times
298	$1.29\text{E}-04$	$3.14\text{E}+00$	$5.05\text{E}+02$
103	$2.41\text{E}-03$	$4.73\text{E}+00$	$5.28\text{E}+02$

Table 8.2 Results for the residual-based and power-based ERBA schemes with g. CPU times are in seconds

$\lvert X_t \rvert$	RMSE	ERBA CPU times	RBA CPU times
82	$1.62\text{E}-01$	$3.63\text{E}+00$	$5.82\text{E}+02$
298	$1.09\text{E}-01$	$4.36\text{E}+00$	$5.40\text{E}+02$

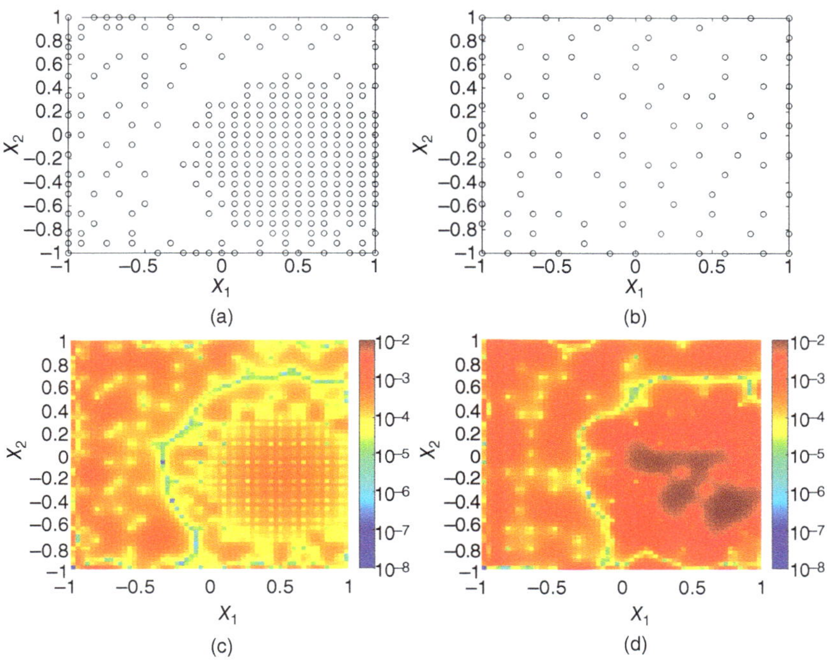

Figure 8.2 Interpolation of f. (a) The reduced residual-based ERBA data set X_t. (b) The reduced power-based ERBA data set X_t. The pointwise relative error between f and the interpolant constructed via the reduced residual-based ERBA (c) and power-based ERBA (d) data set.

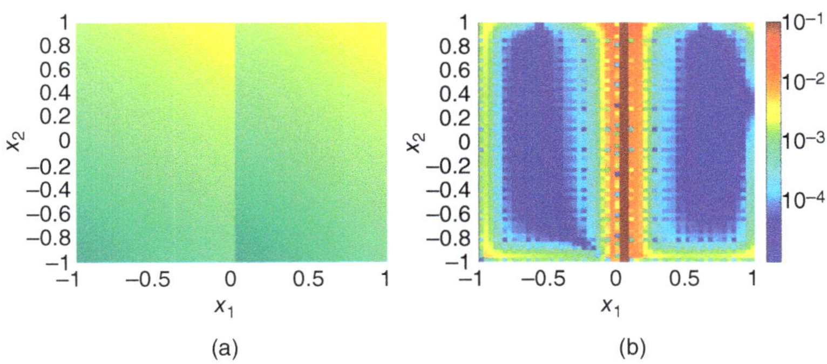

Figure 8.3 (a) The function g. (b) The pointwise relative error between g and the interpolant constructed via the full data set X. The associated RMSE computed on Ξ is $1.14\text{E}-01$.

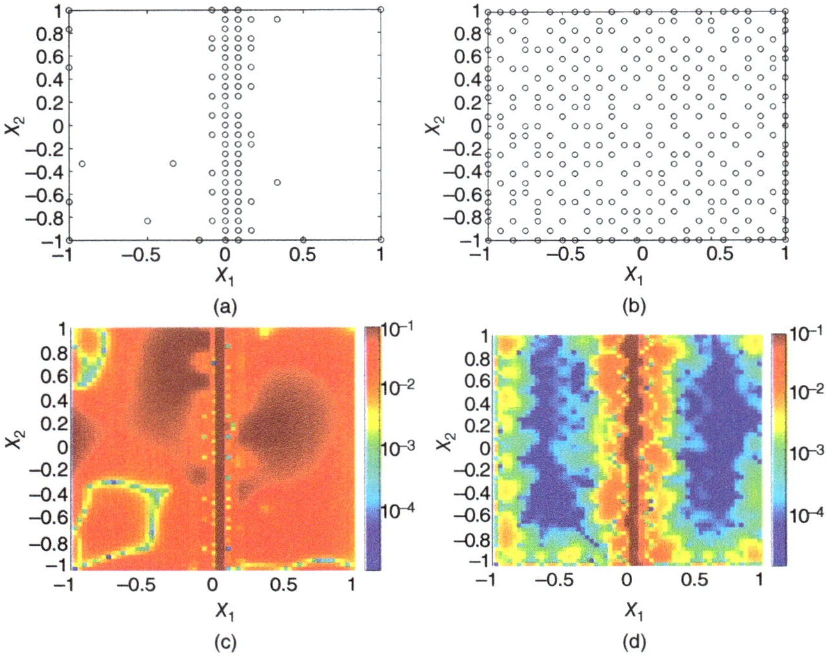

Figure 8.4 Interpolation of g. (a) The reduced residual-based ERBA data set X_t. (b) The reduced power-based ERBA data set X_t. The pointwise relative error between g and the interpolant constructed via the reduced residual-based ERBA (c) and power-based ERBA (d) data set.

9

Approximating via Variably Scaled Discontinuous Kernels

In order to overcome the tuning of the shape parameter and to address the ill-conditioning of kernel matrices, VSKs have been introduced in the context of approximation theory in Bozzini et al. (2015). The underlying idea is to replace the tuning of the shape parameter by treating a *scaling function* as an additional coordinate of the kernel, thereby preserving the advantages of kernel-based interpolation while introducing greater flexibility in the reconstruction process. In this chapter and in the next, we will show how they can outperform standard interpolation techniques.

9.1 Variably Scaled Kernels

We start with the following definition.

Definition 9.1.1 Let $\kappa : \mathbb{R}^{(d+1)\times(d+1)} \longrightarrow \mathbb{R}$ be a continuous strictly positive definite radial kernel and let $\psi : \mathbb{R}^d \longrightarrow \mathbb{R}$ be a scaling function. A VSK κ^{Ψ} on $\mathbb{R}^{d\times d}$ is defined as

$$\kappa^{\Psi}(\boldsymbol{x},\boldsymbol{y}) := \kappa((\boldsymbol{x},\psi(\boldsymbol{x})),(\boldsymbol{y},\psi(\boldsymbol{y}))), \tag{9.1}$$

for $\boldsymbol{x},\boldsymbol{y} \in \mathbb{R}^d$.

We observe that this is a particular case of the construction strategy presented in Proposition 12.2.1 (v). Moreover, we stress that the shape parameter of the underlying kernel κ in Definition 9.1.1 is set to 1, as done in Bozzini et al. (2015, Definition 2.1). We do not include this detail in the notation for the sake of simplicity.

Remark 9.1.1 *Given that we consider radial kernels, the VSK can also be* *defined by writing*

$$\kappa^{\Psi}(\boldsymbol{x},\boldsymbol{y}) = \varphi\left(\sqrt{\|\boldsymbol{x}-\boldsymbol{y}\|^2+|\psi(\boldsymbol{x})-\psi(\boldsymbol{y})|^2}\right),$$

where φ is the usual univariate function related to κ. In other words, if κ is radial on $\mathbb{R}^{(d+1)\times(d+1)}$ so is κ^{Ψ} on $\mathbb{R}^{d\times d}$. Moreover, if κ is (strictly) positive definite so is κ^{Ψ}.

Considering the map $\Psi(\boldsymbol{x}) = (\boldsymbol{x}, \psi(\boldsymbol{x}))$ on Ω and by analogy with (2.2), we can express the interpolant at the set of nodes $\Psi(X) := \{(\boldsymbol{x}_k, \psi(\boldsymbol{x}_k)),\ \boldsymbol{x}_k \in X\}$ as

$$s(\Psi(\boldsymbol{x})) = \sum_{k=1}^{N} c_k \kappa(\Psi(\boldsymbol{x}), \Psi(\boldsymbol{x}_k)), \tag{9.2}$$

with $\boldsymbol{x} \in \Omega$, $\boldsymbol{x}_k \in X$. Therefore, in order to obtain a VSK interpolant s^{Ψ} at the set of nodes X on Ω, it is sufficient to project back the interpolant in (9.2), that is,

$$s^{\Psi}(\boldsymbol{x}) := \sum_{k=1}^{N} c_k \kappa^{\Psi}(\boldsymbol{x},\boldsymbol{x}_k) = \sum_{k=1}^{N} c_k \kappa(\Psi(\boldsymbol{x}), \Psi(\boldsymbol{x}_k)) = s(\Psi(\boldsymbol{x})).$$

An important consequence of this construction is that the analysis of the variably scaled setting is fully understood in terms of the analysis of the underlying standard kernel.

In Proposition 9.1.1, we recall some additional properties of VSKs studied in Bozzini et al. (2015).

Proposition 9.1.1 *Let κ, κ^{Ψ}, and ψ be as in Definition 9.1.1. Then:*

(i) If κ and ψ are continuous, so is κ^{Ψ}.
(ii) According to Definition 4.1.2, we have

$$q_X \le q_{\Psi(X)}$$

for any choice of the scaling map ψ.
(iii) Let $G_{\Psi}(\Omega) = \{(\boldsymbol{x}, \psi(\boldsymbol{x}))\,|\,\boldsymbol{x} \in \Omega\} \subset \Omega \times \mathbb{R}$ be the graph of the scaling function ψ. Then, the native spaces $\mathcal{N}_{\kappa}(G_{\Psi}(\Omega))$ and $\mathcal{N}_{\kappa^{\Psi}}(\Omega)$ are isometrically isomorphic.

Remark 9.1.2 *As a consequence of Proposition 9.1.1 (ii), the variably scaled setting might improve the stability of the interpolation process. For the Gaussian kernel, we prove this intuition in Proposition 16.3.2 in Section 16.3.2.*

The introduction of a scaling function allows us to avoid the tuning of the shape parameter, as we discussed. On the other hand, it is necessary to correctly choose the scaling function to ensure the effectiveness of the variably scaled setting.

In the seminal paper Bozzini et al. (2015), VSKs have been successfully applied in two main directions.

(i) Improving the stability of the interpolation process.
(ii) Enhancing the quality of recovery for the underlying function.

Concerning the former, a possible approach is to use the scaling map to better separate nodes that are too close together, which could lead to instability issues. In a more global sense, one can adopt the variably scaled setting to map the original set of nodes to better data configurations, for example, as in De Marchi et al. (2005). The strength of these approaches is that they depend only on the data sites; thus, they are easily applicable.

On the other hand, the latter approach usually relies on prior knowledge of the underlying function to be interpolated. For example, one can accumulate nodes where the function changes more quickly, or choose the scaling function so that it mimics the underlying function in critical regions (see, e.g., Rossini, 2018).

In Section 9.2 and Chapter 10, we follow this direction and show how the variably scaled setting improves the reconstruction of discontinuous functions with a proper choice of the scaling map.

9.2 Variably Scaled Discontinuous Kernels

We introduce *VSDKs* as a particular case of the VSK setting, where a discontinuous scaling function is used. As a direct consequence, we show that the resulting kernel basis functions are discontinuous.

The purpose of constructing discontinuous kernels is to handle underlying functions that exhibit jumps. Indeed, in this case, standard reconstruction schemes that adopt continuous basis functions provide interpolants that are likely to be affected by the well-known *Gibbs phenomenon*, which causes distortions and artifacts not just in the proximity of the discontinuity points, but also on the whole domain (Fornberg and Flyer, 2008). With a proper choice of the discontinuous scaling map, VSDKs provide interpolants that are robust against the Gibbs phenomenon, as shown in Section 9.2.5.

We first present VSDKs in the univariate case to become acquainted with this framework. After that, we extend the idea to the multivariate setting, providing an accurate characterization of the native spaces of specific discontinuous kernels. For more details, see the papers De Marchi et al. (2020e,b).

9.2.1 Learning from the 1D Case

In the discussion of the univariate case, we assume $\Omega = (a,b) \subset \mathbb{R}$ to be an open interval. Moreover, suppose that we want to interpolate a function $f : \Omega \longrightarrow \mathbb{R}$ that presents ℓ jumps, $\ell \in \mathbb{N}$. More precisely, these jumps are located at the set $D = \{\xi_k,\ k = 1,\dots,\ell\} \subset \Omega$, where $\xi_k < \xi_{k+1}$ for every $k = 1,\dots,\ell-1$.

In this setting, a suitable choice for the scaling map is a function $\psi : \Omega \longrightarrow \mathbb{R}$ that is in turn discontinuous at the points in D. The simplest option is to consider a piecewise constant scaling map ψ alternating between two fixed values.

We point out that, provided these assumptions for the discontinuities, other choices of ψ guarantee similar stable and accurate results (see Section 9.2.5 for more details in the multidimensional framework).
Therefore, we consider piecewise constant scaling functions, such as

$$\psi(x) := \begin{cases} \alpha, & x \in (a,\xi_1) \text{ or } x \in [\xi_k,\xi_{k+1}), \quad \text{where } k \text{ is even}, \\ \beta, & x \in [\xi_k,\xi_{k+1}), \quad \text{where } k \text{ is odd}, \end{cases}$$

and

$$\psi(x)|_{[\xi_\ell,b)} := \begin{cases} \alpha, & \ell \text{ is even}, \\ \beta, & \ell \text{ is odd}. \end{cases}$$

Referring to (9.1) and with this choice of ψ, we call the kernel κ^Ψ a VSDK on $\Omega \times \Omega$, where $\Psi(x) = (x, \psi(x))$, as already defined.

We first present some theoretical results on one dimensional VSDKs. Since the classical theory of kernel-based interpolation presented in Chapter 2 focuses on continuous kernels, it is first convenient to set a VSDK as the limit of a sequence of continuous VSKs. In this way, the discontinuous case naturally inherits some properties of the continuous setting.

The first step consists in expressing the discontinuous scaling map ψ as the pointwise limit of continuous maps. Letting $n \in \mathbb{N}$, $n \geq 1$, we define ψ_n :

$\Omega \longrightarrow \mathbb{R}$ as

$$\psi_n(x) := \begin{cases} \alpha, & x \in (a, \xi_1 - 1/n), \text{ or } x \in [\xi_k + 1/n, \xi_{k+1} - 1/n), \quad k \text{ is even}, \\ \beta, & x \in [\xi_k + 1/n, \xi_{k+1} - 1/n) \quad k \text{ is odd}, \\ \gamma_1(x), & x \in [\xi_k - 1/n, \xi_k + 1/n), \quad k \text{ is odd}, \\ \gamma_2(x), & x \in [\xi_k - 1/n, \xi_k + 1/n), \quad k \text{ is even}, \end{cases}$$

$$\psi_n(x)|_{[\xi_\ell + 1/n, b)} := \begin{cases} \alpha, & \ell \text{ is even}, \\ \beta, & \ell \text{ is odd}, \end{cases}$$

where γ_1, γ_2 are continuous, strictly monotone functions so that

$$\lim_{x \to \xi_{k+1} + 1/n} \gamma_1(x) = \gamma_2(\xi_k - 1/n) = \beta, \qquad \lim_{x \to \xi_{k+1} + 1/n} \gamma_2(x) = \gamma_1(\xi_k - 1/n) = \alpha.$$

It is straightforward to verify that we have the required convergence result

$$\lim_{n \to \infty} \psi_n(x) = \psi(x)$$

for every $x \in \Omega$. We set $\Psi_n(x) = (x, \psi_n(x))$.

The following results are direct consequences of the pointwise convergence of the sequence of continuous scaling maps De Marchi et al. (2020b).

Theorem 9.2.1 *For every $x, y \in \Omega$, we have*

$$\lim_{n \to \infty} \kappa^{\Psi_n}(x, y) = \kappa^{\Psi}(x, y),$$

where κ^{Ψ} is the presented VSDK kernel.

Proof. We can write

$$\lim_{n \to \infty} \kappa^{\Psi_n}(x, y) = \lim_{n \to \infty} \kappa(\Psi_n(x), \Psi_n(y)) = \lim_{n \to \infty} \varphi(\|\Psi_n(x) - \Psi_n(y)\|).$$

Therefore,

$$\begin{aligned} \lim_{n \to \infty} \varphi(\|\Psi_n(x) - \Psi_n(y)\|) &= \varphi\big(\lim_{n \to \infty} \|\Psi_n(x) - \Psi_n(y)\|\big) \\ &= \varphi(\|\Psi(x) - \Psi(y)\|) \\ &= \kappa^{\Psi}(x, y). \end{aligned}$$

□

Corollary 9.2.1 *If $\mathcal{N}_{\kappa^{\Psi_n}}(\Omega)$ is the native space related to κ^{Ψ_n}, taking the limit of the basis functions we obtain the native space $\mathcal{N}_{\kappa^{\Psi}}(\Omega)$ related to κ^{Ψ_n}.*

Proof. If $f \in \mathcal{N}_{\kappa^{\Psi_n}}(\Omega)$, then it can be expressed as a linear combination of basis functions $\kappa^{\Psi}(\cdot,x)$, $x \in \Omega$. From Theorem 9.2.1, we get that for every $x \in \Omega$

$$\lim_{n \to \infty} \kappa^{\Psi_n}(\cdot,x) = \kappa^{\Psi}(\cdot,x),$$

and so f is also a linear combination of the functions $\lim_{n\to\infty} \kappa^{\Psi_n}(\cdot,x)$, $x \in \Omega$. □

Corollary 9.2.2 *Let $f : \Omega \longrightarrow \mathbb{R}$ be a discontinuous function whose discontinuities are located at the points belonging to D. Moreover, let ψ_n and ψ be as in Theorem 9.2.1. Then, considering the interpolation problem (see Problem 2.1.1) with nodes $X = \{x_k,\ k = 1,\dots,N\}$ on Ω, we have*

$$\lim_{n \to \infty} s^{\Psi_n}(x) = s^{\Psi}(x),$$

for every $x \in \Omega$.

Proof. Since s^{Ψ} is a linear combination of the basis functions, the thesis follows from Theorem 9.2.1 and Corollary 9.2.1. □

Recalling the definition of the power function given in (4.3), Theorem 9.2.1 and Corollary 9.2.1 imply that $P_{X,\kappa^{\Psi}}(x) = \lim_{n\to\infty} P_{X,\kappa^{\Psi_n}}\ \forall x \in \Omega$.

Therefore, we can state the following error bound for the VSDK interpolant.

Proposition 9.2.1 *In the assumptions of Corollary 9.2.2, for all $f \in \mathcal{N}_{\kappa^{\Psi}}(\Omega)$ we have*

$$|f(x) - s^{\Psi}(x)| \le P_{X,\kappa^{\Psi}}(x)\|f\|_{\mathcal{N}_{\kappa^{\Psi}}(\Omega)}, \quad x \in \Omega.$$

Proof. We apply the estimate in Theorem 4.1.2 considering the continuous VSK κ^{Ψ_n}. Then, taking the limit $n \to \infty$ and recalling the results of this subsection, the thesis follows. □

We point out that the VSDK interpolant is a linear combination of basis functions that are discontinuous at the jump points of the underlying function to be reconstructed, a significant advantage over the standard basis.

Remark 9.2.1 *We observe that $P_{X,\kappa^{\Psi}}(x) \le P_{X,\kappa}(x)$, $x \in \Omega$, is not guaranteed. Indeed, as shown in De Marchi et al. (2020b), the power function related to the VSDK interpolant could be larger than the standard one, even if s^{Ψ} is more accurate than r.*

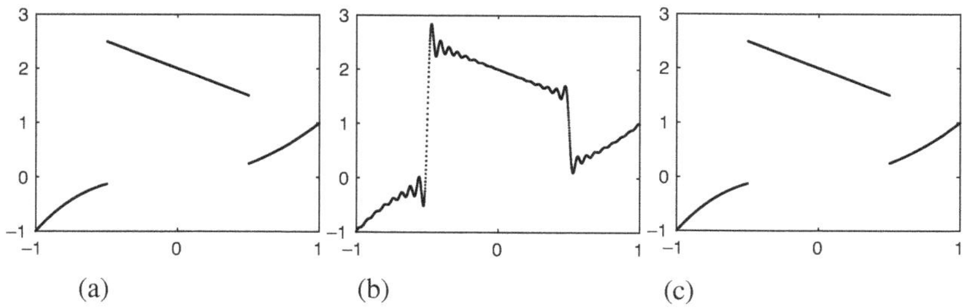

Figure 9.1 Standard and VSDK interpolation of the function f_1. (a) Function f_1; (b) RBF interpolant s_1; (c) VSDK interpolant s_1^{Ψ}.

9.2.1.1 A Numerical Example

We present a numerical example concerning the univariate setting, refer to De Marchi et al. (2020b) for other examples. Let $\Omega = (-1,1)$ and let

$$f_1 : \Omega \longrightarrow \mathbb{R}, \quad f_1(x) = \begin{cases} x^3, & -1 < x < -0.5, \\ -x+2, & -0.5 \le x < 0.5, \\ x^2, & 0.5 \le x < 1, \end{cases}$$

We aim at interpolating f_1 at the set of nodes

$$X = \{x_k = -1 + (k-1)/33,\ k = 1, \ldots, 67\}.$$

We consider a grid of equispaced points on Ω with step size 1.00E-3 as the evaluation set. The a basis function for the RBF interpolation, is the Gaussian function φ_G defined in (4.4).

First, we interpolate the function f_1 via classical RBF interpolation at X, tuning the shape parameter via LOOCV (see Section 5.2). Then, observing that f_1 is discontinuous at $\xi_1 = -0.5$ and $\xi_2 = 0.5$, we consider the VSDK with the scaling map

$$\psi(x) = \begin{cases} 0, & x \in (-1, \xi_1) \text{ or } x \in [\xi_2, 1), \\ 1, & x \in [\xi_1, \xi_2). \end{cases}$$

We denote by s_1 the interpolant obtained by standard RBF interpolation and by s_1^{Ψ} the VSDK interpolant. In Figure 9.1, we show the underlying function and the interpolants obtained. We observe that the standard interpolant is severely affected by the Gibbs phenomenon, while the VSDK interpolant is more accurate. This is numerically confirmed, as shown in Table 9.1.

Table 9.1 Results of standard and VSDK interpolation of f_1

	RBF	VSDK
MAE	1.32E+00	1.42E−04
RMSE	1.16E−01	3.90E−05

9.2.2 Multidimensional VSDKs

To extend the analysis to the multidimensional setting, the first step is to define a multivariate scaling function that replicates the properties of the univariate map.

Assumption 9.2.1 *Let $\Omega \subset \mathbb{R}^d$ be a bounded set, such that*

(i) Ω is the union of m pairwise disjoint sets Ω_k, $k \in \{1,\dots,m\}$.
(ii) Each subset Ω_k, $k = 1,\dots,m$, has a Lipschitz boundary.

Then, let $\Sigma = \{\alpha_1,\dots,\alpha_d\}$, $\alpha_i \in \mathbb{R}$. The scaling function $\psi : \Omega \longrightarrow \Sigma$ is constructed so that

(iii) ψ is piecewise constant, such that $\psi(\boldsymbol{x})|_{\Omega_k} = \alpha_k$.
(iv) The discontinuities of ψ appear only at the boundaries of the subsets Ω_k. Moreover, $\alpha_i \neq \alpha_j$ if Ω_i and Ω_j are neighboring sets.

Taking the so-constructed scaling map ψ, we consider the VSDK $\kappa^{\Psi}(\Omega)$. We aim to reconstruct a function $f : \Omega \longrightarrow \mathbb{R}$ whose discontinuity points are contained in the union of the boundaries of the subsets Ω_k, $k = 1,\dots,m$.

The Lipschitz condition in Assumption 9.2.1 (ii) is crucial. Indeed, it implies an interior cone condition for the subdomains Ω_k (see Adams and Fournier, 2003, p. 84) that plays a fundamental role in the theoretical analysis in Section 9.2.4.

Similarly to the univariate case, if $\boldsymbol{x}$ and $\boldsymbol{y}$ are contained in the same subset $\Omega_k \subset \Omega$ then $\kappa^{\Psi}(\boldsymbol{x},\boldsymbol{y}) = \kappa(\boldsymbol{x},\boldsymbol{y})$.

Remark 9.2.2 *In the hypotheses of Assumption 9.2.1, the same analysis of the univariate case can be repeated. More precisely, since the discontinuous scaling map in Assumption 9.2.1 can be obtained as a pointwise limit of continuous maps, the multivariate VSDK interpolant s^{Ψ} is well-defined in the native space $\mathcal{N}_{\kappa^{\Psi}}(\Omega)$. Therefore, the error bound in Theorem 9.2.1 holds also in the multidimensional case De Marchi et al. (2020b).*

9.2.3 Connections with the Fake Nodes Approach

The interpolation with VSDKs can be considered within the framework of a recent interpolation approach introduced in De Marchi et al. (2020d) for univariate polynomial interpolation. This method, which is called *mapped bases* or Fake Nodes Approach (FNA), has been also extended and applied to barycentric and rational interpolation (see Berrut et al., 2020) and to the multidimensional setting with general bases (see De Marchi et al., 2020c,a).
We now briefly recall the main features of this approach and highlight its connections with VSDKs.

Let us consider Problem 2.1.1 of Section 2.1. Let $M : \Omega \longrightarrow \mathbb{R}^d$ be an injective map and $\{b_1^M, \dots, b_N^M\}$ a related mapped basis, where $b_k^M := b_k \circ M$ for every $k = 1, \dots, N$. Suppose that the set of nodes X is unisolvent for the function space $\mathscr{B}_N^M := \mathrm{span}\{b_1^M, \dots, b_N^M\}$, which is equivalent to the unisolvency of the set $M(X) := \{M(\boldsymbol{x}_1) \dots, M(\boldsymbol{x}_N)\}$ with respect to the space $\mathscr{B}_N$ (see Definition 2.1.2).
Then, we can define the interpolant s^M at X on Ω as

$$s^M(\boldsymbol{x}) := s(M(\boldsymbol{x})) = \sum_{k=1}^{N} c_k b_i(M(\boldsymbol{x})).$$

Comparing the VSDK interpolant s^Ψ with the fake nodes interpolant s^M, we first observe that the maps Ψ and M play a similar role: indeed, both determine the construction of a new interpolant out of a standard scheme by previously mapping the original set of nodes.

The main difference between the two approaches is that, in the former, the dataset is mapped to an upper-dimensional space. In contrast, in the latter, the dimension of the original interpolation domain is preserved. We point out that kernel-based interpolation is generally blind to the dimensional augmentation of the space, whereas this can represent a problem when considering other bases (e.g., in the polynomial setting).

Let $\alpha_0 \in \mathbb{R}$ and consider the FNA in the RBF interpolation setting. Then, let

$$\mathsf{D}_{ij}^\Psi = \|\Psi(\boldsymbol{x}_i) - \Psi(\boldsymbol{x}_j)\|, \quad \mathsf{D}_{ij}^M = \|M(\boldsymbol{x}_i) - M(\boldsymbol{x}_j)\|$$

be the distance matrices related to the VSDK scheme and to the FNA. As explained in De Marchi et al. (2020c), and according to Assumption 9.2.1, if the map M is defined as

$$M(\boldsymbol{x}) = \boldsymbol{x} + \sum_{k=1}^{m} \boldsymbol{\alpha}_k \chi_{\Omega_k}(\boldsymbol{x}),$$

where χ_{Ω_k} is the characteristic function corresponding to Ω_k and $\boldsymbol{\alpha}_k = (k\alpha_0, k\alpha_0)$, then D^{Ψ} is *similar* to D^{M} in the following sense: since they both preserve the distances between nodes from the same subregion and similarly enlarge both distances between nodes lying in different subregions (see Example 9.2.1), very often the resulting distance matrices do not differ significantly, and the obtained interpolants s^{Ψ} and s^{M} provide similar accuracy results.

Example 9.2.1 *In Figure 9.2, a set of equispaced nodes in $\mathbb{R}$ (a) is mapped to $\mathbb{R}^2$ in the VSDKs setting (b), while it is shifted in $\mathbb{R}$ considering the FNA (c).*

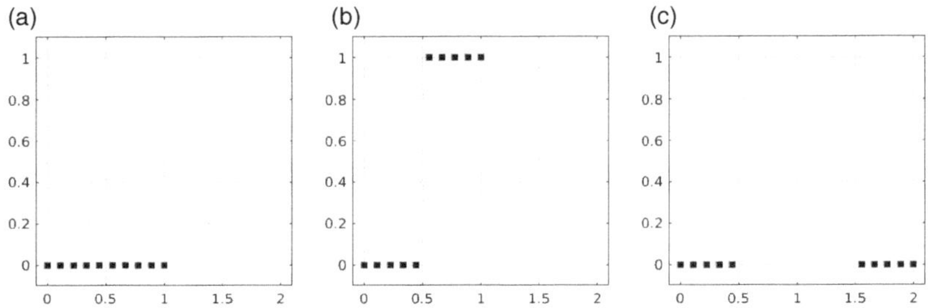

Figure 9.2 (a–c) The difference between the two discussed strategies, VSDKs and FNA.

9.2.4 Error Estimates and Native Spaces for VSDKs

In this subsection, we investigate the theoretical properties of native spaces $\mathcal{N}_{\kappa^{\Psi}}(\Omega)$, whose underlying standard kernel κ is such that its related univariate radial function φ has the following Fourier decay

$$(\mathrm{F}\varphi)(\boldsymbol{\omega}) \sim (1+\|\boldsymbol{\omega}\|^2)^{-\beta-\frac{1}{2}}, \quad \beta > \frac{d-1}{2}. \tag{9.3}$$

As presented in (4.7) in Section 4.2, these are Matérn functions, whose native space is the Sobolev space $\mathscr{W}_2^{\beta+\frac{1}{2}}(\Omega)$. The motivation behind such a choice of the value of the regularity parameter will be clear after Theorem 9.2.2.

From now on, we use the notation $\tilde{\boldsymbol{x}} = \Psi(\boldsymbol{x})$ for the elements of the graph $G_{\Psi}(\Omega)$.

Let $\beta \geq 0$ and $1 \leq p \leq \infty$. Taking into account the subdivision of the domain Ω described in Assumption 9.2.1, we define the space

$$\mathscr{W}\mathscr{P}_p^{\beta}(\Omega) := \left\{ w : \Omega \longrightarrow \mathbb{R} \,|\, w|_{\Omega_k} \in \mathscr{W}_p^{\beta}(\Omega_k),\ k \in \{1,\dots,m\} \right\}.$$

This space contains the piecewise smooth functions w on Ω whose restriction to any subregion Ω_k, $i = k,\dots,m$ is contained in the standard Sobolev space

$\mathscr{W}_p^\beta(\Omega_k)$. Moreover, we endow the space $\mathscr{WP}_p^\beta(\Omega)$ with the norm

$$\|w\|^p_{\mathscr{WP}_p^\beta(\Omega)} = \sum_{k=1}^m \|w|_{\Omega_k}\|^p_{\mathscr{W}_p^\beta(\Omega_k)}.$$

Remark 9.2.3 *For any subdivision of Ω satisfying Assumption 9.2.1, the standard Sobolev space $\mathscr{W}_p^\beta(\Omega)$ is contained in $\mathscr{WP}_p^\beta(\Omega)$. Moreover, $\mathscr{WP}_p^\beta(\Omega)$ and the related norm are strongly dependent on the chosen subdivision.*

Here, we refer to (De Marchi et al., 2020e, Lemma 3.1), since it introduces some essential facts that are relevant to characterize the native space $\mathscr{N}_{\kappa^\Psi}(\Omega)$. See De Marchi et al. (2020e) for a detailed proof.

Lemma 9.2.1 *We have the following relations for the extension and trace operators (see Wendland, 2005, Theorem 10.46 and Theorem 10.47) in the native spaces (i) and Sobolev spaces (ii):*

(i) Every $f \in \mathscr{N}_\kappa(G_\Psi(\Omega))$ has a natural extension $\mathrm{E}f \in \mathscr{N}_\kappa(\mathbb{R}^{d+1})$. In addition,

$$\|\mathrm{E}f\|_{\mathscr{N}_\kappa(\mathbb{R}^{d+1})} = \|f\|_{\mathscr{N}_\kappa(G_\Psi(\Omega))}.$$

For every $g \in \mathscr{N}_\kappa(\mathbb{R}^{d+1})$, the trace $\mathrm{Tr}_{G_\Psi(\Omega)}g$ is contained in $\mathscr{N}_\kappa(G_\Psi(\Omega))$. Further,

$$\|\mathrm{Tr}_{G_\Psi(\Omega)}g\|_{\mathscr{N}_\kappa(G_\Psi(\Omega))} \leq \|g\|_{\mathscr{N}_\kappa(\mathbb{R}^{d+1})}.$$

(ii) Let $\beta > \frac{1}{2}$. For every $g \in \mathscr{W}_2^\beta(\mathbb{R}^{d+1})$ the trace $\mathrm{Tr}_{G_\Psi(\Omega)}g$ is contained in $\mathscr{W}_2^{\beta-1/2}(G_\Psi(\Omega))$ and the trace operator $\mathrm{Tr}_{G_\Psi(\Omega)} : \mathscr{W}_2^\beta(\mathbb{R}^{d+1}) \longrightarrow \mathscr{W}_2^{\beta-1/2}(G_\Psi(\Omega))$ is bounded.
Furthermore, there exists a bounded extension operator $\mathrm{E} : \mathscr{W}_2^{\beta-1/2}(G_\Psi(\Omega)) \longrightarrow \mathscr{W}_2^\beta(\mathbb{R}^{d+1})$ such that $\mathrm{Tr}_{G_\Psi(\Omega)}\mathrm{E}f = f$ for all $f \in \mathscr{W}_2^{\beta-1/2}(G_\Psi(\Omega))$.

We are ready to state the following characterization (see De Marchi et al., 2020e, Theorem 3.2).

Theorem 9.2.2 *Let Assumption 9.2.1 hold. Moreover, assume that the continuous strictly positive definite radial kernel $\kappa : \mathbb{R}^{d+1} \times \mathbb{R}^{d+1} \to \mathbb{R}$, based on the radial function φ, satisfies the decay condition in* (9.3).
Then, for the discontinuous kernel κ^Ψ, we have

$$\mathscr{N}_{\kappa^\Psi}(\Omega) = \mathscr{WP}_2^\beta(\Omega),$$

with the norms of the two Hilbert spaces being equivalent.

Proof. By means of the results in Lemma 9.2.1, it is possible to construct the following forward and backward chain of Hilbert space operators:

$$\mathcal{N}_{\kappa^{\Psi}}(\Omega) \underset{\mathrm{Q}_1}{\overset{\mathrm{P}_5}{\leftrightarrows}} \mathcal{N}_{\kappa}(G_{\Psi}(\Omega)) \underset{\mathrm{Q}_2}{\overset{\mathrm{P}_4}{\leftrightarrows}} \mathcal{N}_{\kappa}(\mathbb{R}^{d+1}) \underset{\mathrm{Q}_3}{\overset{\mathrm{P}_3}{\leftrightarrows}} \mathcal{W}_2^{\beta+1/2}(\mathbb{R}^{d+1}) \underset{\mathrm{Q}_4}{\overset{\mathrm{P}_2}{\leftrightarrows}} \mathcal{W}_2^{\beta}(G_{\Psi}(\Omega)) \underset{\mathrm{Q}_5}{\overset{\mathrm{P}_1}{\leftrightarrows}} \mathcal{WP}_2^{\beta}(\Omega),$$

which are defined as follows.

$$\begin{aligned}
&\mathrm{P}_1: \quad \mathrm{P}_1 f = \tilde{f}, \quad \text{with} \quad \tilde{f}(\tilde{\boldsymbol{x}}) = f(\boldsymbol{x}), \quad \text{for all } \boldsymbol{x} \in \Omega,\\
&\mathrm{P}_2: \quad \mathrm{P}_2 f = \mathrm{E} f, \quad \text{(Extension in the sense of Lemma 9.2.1 (ii))}\\
&\mathrm{P}_3: \quad \mathrm{P}_3 f = f,\\
&\mathrm{P}_4: \quad \mathrm{P}_4 f = \mathrm{Tr}_{G_{\Psi}(\Omega)} f, \quad \text{(Trace in the sense of Lemma 9.2.1 (i))}\\
&\mathrm{P}_5: \quad \mathrm{P}_5 \tilde{f} = f, \quad \text{with} \quad f(\boldsymbol{x}) = \tilde{f}(\tilde{\boldsymbol{x}}), \quad \text{for all } \boldsymbol{x} \in \Omega\\
&\mathrm{Q}_1: \quad \mathrm{Q}_1 f = \tilde{f}, \quad \text{with} \quad \tilde{f}(\tilde{\boldsymbol{x}}) = f(\boldsymbol{x}), \quad \text{for all } \boldsymbol{x} \in \Omega,\\
&\mathrm{Q}_2: \quad \mathrm{Q}_2 f = \mathrm{E} f, \quad \text{(Extension in the sense of Lemma 9.2.1 (i))}\\
&\mathrm{Q}_3: \quad \mathrm{Q}_3 f = f,\\
&\mathrm{Q}_4: \quad \mathrm{Q}_4 f = \mathrm{Tr}_{G_{\Psi}(\Omega)} f, \quad \text{(Trace in the sense of Lemma 9.2.1 (ii))}\\
&\mathrm{Q}_5: \quad \mathrm{Q}_5 \tilde{f} = f, \quad \text{with} \quad f(\boldsymbol{x}) = \tilde{f}(\tilde{\boldsymbol{x}}), \quad \text{for all } \boldsymbol{x} \in \Omega.
\end{aligned}$$

All these 10 operators are well-defined and continuous: P_5 and Q_1 are isometries by Proposition 9.1.1 (iii). P_2, Q_2 and P_4, Q_4 are continuous by Lemma 9.2.1. Since κ satisfies the condition (9.3), the native space $\mathcal{N}_{\kappa}(\mathbb{R}^{d+1})$ is equivalent to the Sobolev space $\mathcal{W}_2^{\beta}(\mathbb{R}^{d+1})$ (see Wendland, 2005, corollary 10.48). Therefore, also the identity mappings P_3 and Q_3 are continuous. Finally, the Sobolev norm for a function $\tilde{f}$ on the graph $G_{\Psi}(\Omega)$ is a reformulation of the norm of $f \in \mathcal{WP}_2^{\beta}(\Omega)$. Therefore, also the operators P_1 and Q_5 are isometries.

Then, the concatenations $\mathrm{P}_5\mathrm{P}_4\mathrm{P}_3\mathrm{P}_2\mathrm{P}_1$ and $\mathrm{Q}_5\mathrm{Q}_4\mathrm{Q}_3\mathrm{Q}_2\mathrm{Q}_1$ are continuous operators. Since $\mathrm{P}_5\mathrm{P}_4\mathrm{P}_3\mathrm{P}_2\mathrm{P}_1$ is the inverse of $\mathrm{Q}_5\mathrm{Q}_4\mathrm{Q}_3\mathrm{Q}_2\mathrm{Q}_1$, these two operators therefore provide an isomorphism between the Hilbert spaces $\mathcal{N}_{\kappa^{\Psi}}(\Omega)$ and $\mathcal{WP}_2^{\beta}(\Omega)$. □

We now present an error bound for a VSDK interpolant s^{Ψ} that interpolates a function $f \in \mathcal{WP}_2^{\beta}(\Omega)$ at the set of nodes $X = \{\boldsymbol{x}_k,\ k = 1, \dots, N\} \subset \Omega$. First, we introduce the *regional* fill distance (see also Definition 4.1.3)

$$h_k := h_{X,\Omega_k} = \sup_{\boldsymbol{x} \in \Omega_k} \inf_{\boldsymbol{x}_k \in X \cap \Omega_k} \|\boldsymbol{x} - \boldsymbol{x}_k\|$$

and the *global* fill distance

$$h := \max_{k \in \{1,\dots,m\}} h_k$$

Moreover, we need the following well-known Sobolev sampling inequality (see Narcowich et al., 2005, Theorem 2.12).

Proposition 9.2.2 *Let $\beta > 0$, as well as $1 \leq p < \infty$ and $1 \leq q \leq \infty$. Furthermore, let $\gamma \in \mathbb{N}_0$ be such that $\lfloor \beta \rfloor > \gamma + d/p$ (for $p = 1$ also equality is possible) and u be a function that vanishes on $X \cap \Omega_k$. Then, there is a $h_0 > 0$ such that for $h_k \leq h_0$ and for the subregions Ω_k satisfying Assumption 9.2.1 (ii) we have the Sobolev inequality*

$$\|u\|_{\mathscr{W}_q^\gamma(\Omega_k)} \leq C_k h_k^{\beta-\gamma-d(1/p-1/q)_+} \|u\|_{\mathscr{W}_p^\beta(\Omega_k)}.$$

The constant $C_k > 0$ is independent of h_k.

Concerning Proposition 9.2.2, we point out that the value h_0 depends on the Sobolev regularity β and on the interior cone condition of the subsets Ω_k as observed in Narcowich et al. (2005).

We are ready to state the following error bound (De Marchi et al., 2020e, Theorem 3.4).

Theorem 9.2.3 *Let Assumption 9.2.1 be satisfied. Furthermore, let $\beta > 0$, $1 \leq q \leq \infty$, and $\gamma \in \mathbb{N}_0$ be such that $\lfloor \beta \rfloor > \gamma + \frac{d}{2}$. Moreover, suppose that the radial function φ related to the kernel κ satisfies the Fourier decay in* (9.3)*. Then, for $f \in \mathscr{WP}_2^\beta(\Omega)$, we obtain for all $h \leq h_0$ the error estimate*

$$\|f - s^\Psi\|_{\mathscr{WP}_q^\gamma(\Omega)} \leq C h^{\beta-\gamma-d(1/2-1/q)_+} \|f\|_{\mathscr{WP}_2^\beta(\Omega)}, \tag{9.4}$$

where the constant $C > 0$ is independent of h.

Proof. By assumption, the function f is an element of $\mathscr{WP}_2^\beta(\Omega)$. Furthermore, the native space characterization in Theorem 9.2.2 guarantees that the VSDK interpolant s^Ψ is also an element of $\mathscr{WP}_2^\beta(\Omega)$. Therefore, we can apply Proposition 9.2.2 with $p = 2$ to every subset Ω_i and obtain

$$\|f - s^\Psi\|_{\mathscr{W}_q^\gamma(\Omega_i)} \leq C_i h_i^{\beta-\gamma-d(1/2-1/q)_+} \|f - s^\Psi\|_{\mathscr{W}_2^\beta(\Omega_i)}, \quad i \in \{1, \ldots, m\},$$

with $h_i \leq h_0$. Now, using the definition of the piecewise Sobolev space $\mathscr{WP}_q^\gamma(\Omega)$, we can synthesize these estimates to obtain

$$\|f - s^\Psi\|_{\mathscr{WP}_q^\gamma(\Omega)} \leq C_{\max} h^{\beta-\gamma-d(1/2-1/q)_+} \|f - s^\Psi\|_{\mathscr{WP}_2^\beta(\Omega)},$$

where $C_{\max} = \max_{1 \leq i \leq m} C_i$ and $h = \max_{1 \leq i \leq m} h_i$. Since $\mathscr{WP}_2^\beta(\Omega)$ is equivalent to the native space $\mathscr{N}_{\kappa^\Psi}(\Omega)$, we can use the fact that the interpolant s^Ψ is a projection into a subspace of $\mathscr{N}_{\kappa^\Psi}(\Omega)$. This helps us to finalize our bound:

$$\|f - s^\Psi\|_{\mathscr{WP}_2^\beta} \leq C' \|f - s^\Psi\|_{\mathscr{N}_{\kappa^\Psi}(\Omega)} \leq C' \|f\|_{\mathscr{N}_{\kappa^\Psi}(\Omega)} \leq C' C'' \|f\|_{\mathscr{WP}_2^\beta(\Omega)},$$

with the two constants C', C'' describing the bounds for the equivalence of the

two *Hilbert space norms* . The constant C in (9.4) is then given by $C = C'C''C_{\max}$.
□

Remark 9.2.4 *The convergence rate of* s^{Ψ} *toward* $f \in \mathscr{WP}_2^{\beta}(\Omega)$ *depends only on the smoothness* β *of the underlying function in the interior of the subsets of the subdivision. This is a crucial advantage with respect to standard RBF interpolation, since in the VSDK setting, the jumps at the boundaries of the subsets do not interfere with the overall interpolation.*

In Section 9.2.5, numerical tests confirm these theoretical results, with the Gibbs phenomenon that essentially does not appear.

9.2.5 Numerics

In this section, we perform various 2D numerical tests and present, in particular, the results as shown in De Marchi et al. (2020e):

1. First, we present an example considering gridded data, similarly to what we showed in Section 9.2.1 for the univariate case.
2. Then, we define and take the set of *Lissajous nodes* as nodal sets, a set of points used in applications concerning MPI (see Chapter 10). The theoretical results obtained in Section 9.2.4 are confirmed by various tests.
3. Finally, we focus on the scaling function. More precisely, we test the reconstruction process's sensitivity to inaccurate scaling functions, which can affect the overall accuracy of the final reconstruction.

We point out that, the positions of the jumps of the underlying functions are analytically known *a priori*, which allows proper verification of the theoretical results. In Section 10.1, we consider the problem of detecting the edges from the data, which is typical in applications.

The domain of our experiments is $\Omega = [-1,1]^2$, as it is the natural domain of the *Lissajous curves* generating the *Lissajous nodes* Erb et al. (2016). The reconstructions are evaluated on a 150×150 equispaced grid Ξ on Ω.

9.2.5.1 Interpolation at Gridded Data

Let $f_2 : \Omega \longrightarrow \mathbb{R}$ be defined as

$$f_2(x,y) := \begin{cases} x + y^2 + 0.5, & 2x^2 + 3y^2 \leq 0.5, \\ 0.5, & y \geq x^3 + 1, \\ -(x+y)/8, & \text{otherwise.} \end{cases}$$

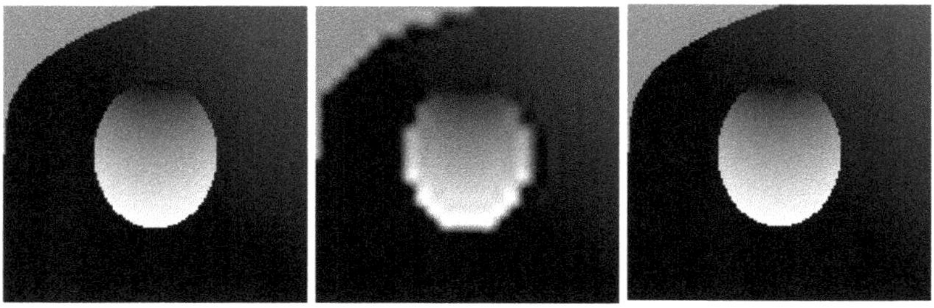

(a) (b) (c)

Figure 9.3 Standard and VSDK interpolation of the function f_2. (a) Function f_2; (b) RBF interpolant s_2; (c) VSDK interpolant s_2^Ψ .

Table 9.2 Results of standard and VSDK interpolation of f_2

	RBF	VSDK
MAE	1.04E+00	1.31E−02
RMSE	7.50E−02	7.27E−04
SSIM	0.861	0.999

As an interpolation set, we consider the equispaced grid in Ω with step $1/13$. In this example, we take the Wendland function $\varphi_{W,2}$ defined in (4.10) as a basis function.

First, we interpolate f_2 using classical RBFs on the chosen interpolation set, tuning the shape parameter via LOOCV. Then, we consider the VSDK setting and we choose the scaling map

$$\psi(x,y) := \begin{cases} 2, & 2x^2+3y^2 \leq 0.5, \\ 1, & y \geq x^3+1, \\ 0, & \text{otherwise.} \end{cases}$$

We denote by s_2 the interpolant obtained by standard RBF interpolation and by s_2^Ψ the VSDK interpolant. In Table 9.2, to verify the quality of the reconstructions, we consider the MAE and RMSE, defined in (6.2), and the SSIM (Structural SIMilarity) index Wang et al. (2004), which is widely used to measure the similarity between two images. More specifically, the SSIM index takes values in the interval $(-1,1]$, with larger values indicating better quality.
We observe that the VSDK interpolant is much more accurate than the standard RBF interpolant. The reconstructions are shown in Figure 9.3.

9.2.5.2 Tests with Lissajous Nodes

Let $\boldsymbol{N} = (N_1, N_2) \in \mathbb{N}^2$ be a vector, where N_1, N_2 are relatively prime. Given $\varepsilon \in \{1,2\}$, the generating curves for the *Lissajous nodes* are defined

$$\gamma_\varepsilon^{\boldsymbol{N}}(t) := \left(\cos(N_2 t), \cos\left(N_1 t - \tfrac{\varepsilon-1}{2N_2}\pi\right)\right).$$

The Lissajous curve $\gamma_\varepsilon^{\boldsymbol{N}}$ is contained in Ω. Moreover, depending on the parameter ε, we obtain two different classes of curves. Indeed, if $\varepsilon = 1$, the curve is degenerate and is π-periodic, while if $\varepsilon = 2$ the curve is nondegenerate and is 2π-periodic.

Remark 9.2.5 *Well-known examples of degenerate curves are the curves with parameters* $\boldsymbol{N} = (N, N+1)$, $N \in \mathbb{N}$, *since they are the generating curves of the Padua points (Bos et al., 2006; Erb, 2016).*

Nondegenerate curves where $N_1 + N_2$ is odd are typically adopted in magnetic particle imaging, MPI (Erb et al., 2016; Kaethner et al., 2016; Knopp and Buzug, 2012) and have already been considered in De Marchi et al. (2017) and Erb et al. (2015). Therefore, in our tests, we restrict ourselves to such curves.

The *Lissajous nodes* sampled from $\gamma_\varepsilon^{\boldsymbol{N}}$ are

$$LS_\varepsilon^{\boldsymbol{N}} := \left\{\gamma_\varepsilon^{\boldsymbol{N}}\left(\frac{\pi k}{\varepsilon N_1 N_2}\right), \quad k = 0, \ldots, 2\varepsilon N_1 N_2 - 1\right\}. \tag{9.5}$$

The nodes $LS_2^{\boldsymbol{N}}$ are illustrated in Figure 9.4 for $\boldsymbol{N} = (33, 32)$ and $\boldsymbol{N} = (10, 11)$.

In our setting, the number of nodes is given by $LS_2^{\boldsymbol{N}} = 2N_1N_2 + N_1 + N_2$ Erb et al. (2016). Using the notation $S_N = \sin(\pi/N)$, the related fill distance $h_{LS_2^{\boldsymbol{N}},\Omega}$ can be expressed as (De Marchi et al., 2020e)

$$h_{LS_2^{\boldsymbol{N}},\Omega} = \frac{1}{2}\max\left\{\sqrt{S_{N_1}^2 + \left(\frac{S_{2N_1}^2 + S_{2N_2}^2 - S_{N_1}S_{2N_1}}{S_{2N_2}}\right)^2}, \sqrt{S_{N_2}^2 + \left(\frac{S_{2N_1}^2 + S_{2N_2}^2 - S_{N_2}S_{2N_2}}{S_{2N_1}}\right)^2}\right\}.$$

Moreover, the following estimates hold

$$\frac{1}{2}\max\{S_{N_1}, S_{N_2}\} \le h_{LS_2^{\boldsymbol{N}}} \le \max\{S_{2N_1}, S_{2N_2}\} \le \max\left\{\frac{\pi}{2N_1}, \frac{\pi}{2N_2}\right\}.$$

Here, to test the VSDK setting, we consider the Shepp–Logan phantom (Shepp and Logan, 1974) f_{SL} on Ω. More precisely, the function $f_{\mathrm{SL}} : \Omega \longrightarrow [0,1]$ consists of a composition of ten-step functions defined by elliptic equations. In Figure 9.4, f_{SL} is evaluated on the equispaced grid Ξ.

For testing the results presented in Section 9.2.4, we consider the Matérn RBFs in (4.9) and the Gaussian in (4.4) for the kernel-based approximation. Indeed, the native space of the Gaussian kernel is contained in every Sobolev space

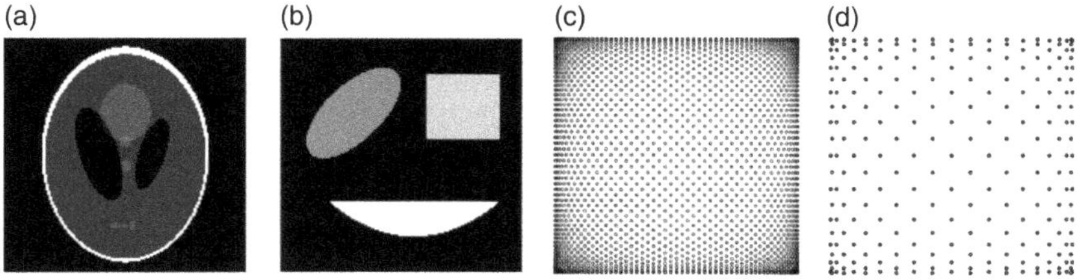

Figure 9.4 The function f_{SL} (a), the function f_{G} (b), the Lissajous nodes $LS_2^{(33,32)}$ (c) and $LS_2^{(10,11)}$ (d).

$\mathscr{W}_2^{\beta}(\mathbb{R}^d)$, $\beta \geq 0$, while the native spaces of the Matérn kernels are reported in Table 9.3.

Table 9.3 Native spaces of Matérn kernels in $\mathbb{R}^d$

RBF	Native spaces
$\varphi_{M,0}$	$\mathscr{W}_2^{(d+1)/2}(\mathbb{R}^{\beta})$
$\varphi_{M,2}$	$\mathscr{W}_2^{(d+3)/2}(\mathbb{R}^{\beta})$
$\varphi_{M,4}$	$\mathscr{W}_2^{(d+5)/2}(\mathbb{R}^{\beta})$

As discussed in Sections 4.1 and 4.2, the decrease in the separation distance of the interpolation nodes could lead to instability issues, especially when considering the Gaussian kernel. Therefore, we stabilize the process by adding multiple $\lambda = 10^{-12}$ of the identity to the interpolation matrix in the linear system in (2.3). This technique, usually referred as *Tikhonov regularization*, is a classic regularization method (see Section 12.1 for more details) that is used to avoid bad conditioning in linear systems (see, e.g., Fasshauer and McCourt, 2015, chapters 11, 12, and 13 for other regularization techniques).

For $N \in \{4, 8, 12, \ldots, 40\}$, we interpolate the Shepp–Logan phantom at the Lissajous nodes $LS_2^{(N+1,N)}$, adopting the Matérn and Gaussian kernels and computing the standard RBF as well as VSDK interpolants for all sets of Lissajous nodes. More precisely, the scaling function for the VSDK interpolant is constructed as follows. Since the function f_{SL} assumes the values $f_i \in \{0, 0.1, 0.2, 0.3, 0.4, 1\}$, $i \in \{0, \ldots, 5\}$, we may consider the piecewise constant scaling map on Ω

$$\psi_\delta(\boldsymbol{x}) = \delta\, i \quad \text{if } f_{\mathrm{SL}}(\boldsymbol{x}) = f_i,$$

where $\delta > 0$ is a separation parameter. We observe that taking $\delta = 0$ we have

$\psi \equiv 0$ and thus the VSDK interpolant reduces to the standard RBF interpolant. All reconstructions are evaluated on the grid Ξ.

We are interested in evaluating the behavior of the $\mathscr{L}_2$-error between the original Shepp–Logan function f_{SL} and the interpolants against the introduced fill distance $h = h_{LS_2^N,\Omega}$. More precisely, we compute the RMSE (see 6.2) to approximate the continuous $\mathscr{L}_2$-error. We point out that, although adopting the fill distance $h_{LS_2^N,\Omega}$ allows a good comparison between the different interpolation schemes, it is generally smaller than the fill distance considered in Theorem 9.2.3, which depends on the decomposition of the domain Ω into the disjoint subsets Ω_i.

In Figure 9.5a, b, we display the log–log diagrams for different VSDK interpolants with $\delta = 0.01$ and for standard RBF interpolants. Moreover, we show the behavior of the interpolation error for different values of δ in Figure 9.5c. Concerning the separation parameter, in Table 9.4 we report the slopes of the error regression lines for $\delta \in \{0, 0.01, 0.1, 1\}$. Finally, in Figure 9.6 we show different reconstructions of the Shepp–Logan phantom at the nodes $LS_2^{(33,32)}$.

Regarding the conducted tests, we first observe that the VSDK scheme is overall more accurate than the standard RBF reconstruction algorithm. For all the kernels considered, the slope of the regression line for the standard RBF interpolation ($\delta = 0$) is between 0.12 and 0.25 (see Figure 9.5). This is due to the low order of smoothness of the Shepp-Logan function, which causes the appearance of the Gibbs phenomenon.

On the other hand, if $\delta > 0$ then the Shepp–Logan phantom is contained in all piecewise Sobolev spaces $\mathscr{WP}_2^{\beta}(\Omega)$, $s \geq 0$, thus the theoretical estimates in Theorem 9.2.3 have to be considered. More precisely, the predicted theoretical convergence rate for the $\mathscr{L}_2$-error is in order of the smoothness β of the applied kernel. Indeed, the influence of β is visible in Table 9.4 and Figure 9.5.

The impact of the scaling function's separation parameter, δ, is also notable. At first glance, this dependence of the VSDK interpolation on δ is not visible in Theorem 9.2.3. Nevertheless, the constant C in the error estimate inherently depends on the scaling function ψ_δ, since the proof of Theorem 9.2.3 relies on the equivalence of the native space in Theorem 9.2.2 De Marchi et al. (2020e). Starting from a small value of δ, the accuracy of the reconstruction improves as such a value increases. When $\delta \gtrsim 10$, the resulting VSDK interpolant stops depending on the parameter δ.

9.2.5.3 Perturbations of the Scaling Map

Let $\boldsymbol{x} = (x_1, x_2) \in \Omega$. We consider the geometric composition of:

(i) the ellipse $E = \{\boldsymbol{x} \mid 2(x_2 + x_1 + 0.05)^2 + 9(x_2 - x_1 - 0.75)^2 \leq 1\}$;

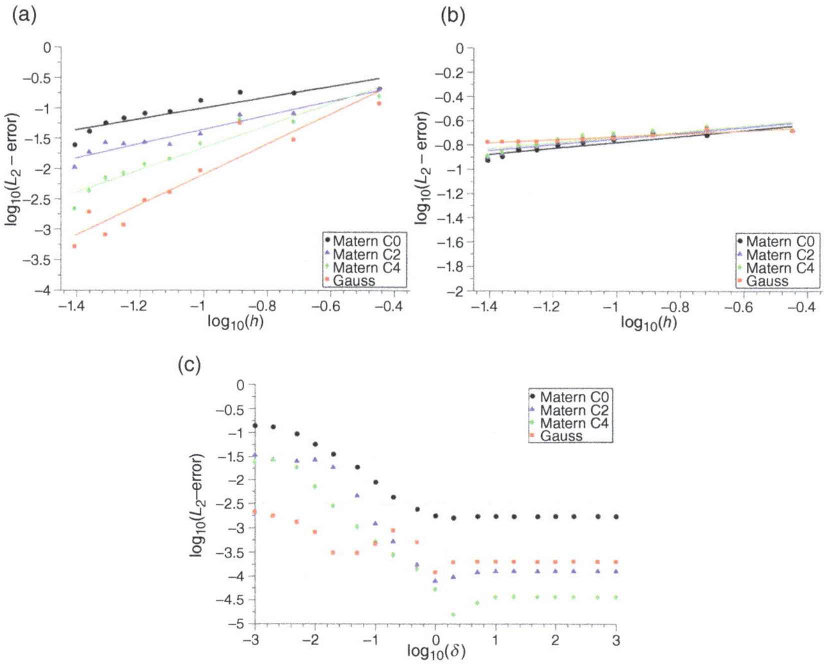

Figure 9.5 Convergence rates for interpolating the function f_{SL} at the nodes $LS_2^{\boldsymbol{N}}$: adopting the VSDK scheme with scaling function ψ_δ, $\delta = 0.01$ (a), and using the standard RBF scheme (b). Furthermore, (c), the $\mathscr{L}_2$-error for interpolation at the nodes $LS_2^{(33,32)}$ varying δ is displayed.

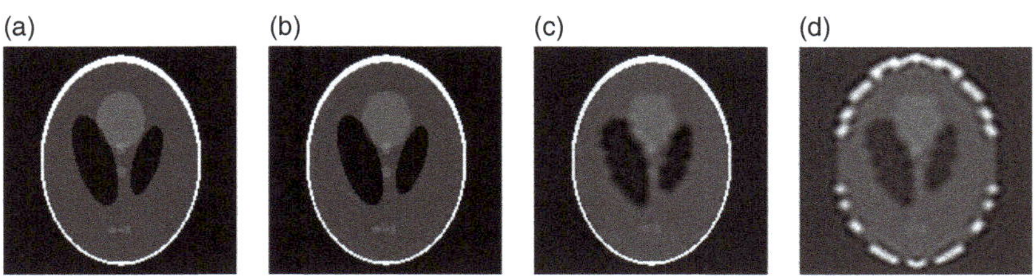

Figure 9.6 Different reconstructions of the Shepp–Logan phantom at the nodes $LS_2^{(33,32)}$ using the radial function $\varphi_{M,2}$. (a–d) the results adopting the VSDK scheme with scaling function ψ_δ, $\delta \in \{1, 0.1, 0.01\}$ and the standard RBF interpolation ($\delta = 0$).

(ii) the rectangle $R = \{\boldsymbol{x} \mid |x_1 - 0.5| \leq 0.3, |x_2 - 0.3| \leq 0.28\}$;

(iii) the parabola $P = \{\boldsymbol{x} \mid x_2 \leq -0.4,\ x_2 + 0.7 \geq 0.6(x_1 - 0.1)^2\}$.

Table 9.4 For the VSDK and RBF schemes, the slopes of the regression lines for the computed logarithmic $\mathscr{L}_2$-errors

Kernel	Theoretical slope	Experimental slope			
		$\delta = 1$	$\delta = 0.1$	$\delta = 0.01$	$\delta = 0$
$\varphi_{M,0}$	1.5	2.2569	1.9131	0.8933	0.2493
$\varphi_{M,2}$	2.5	2.9859	2.5753	1.1715	0.2403
$\varphi_{M,4}$	3.5	3.1380	2.5331	1.8237	0.2355
φ_G	Analytic	3.3858	2.8668	2.4959	0.1260

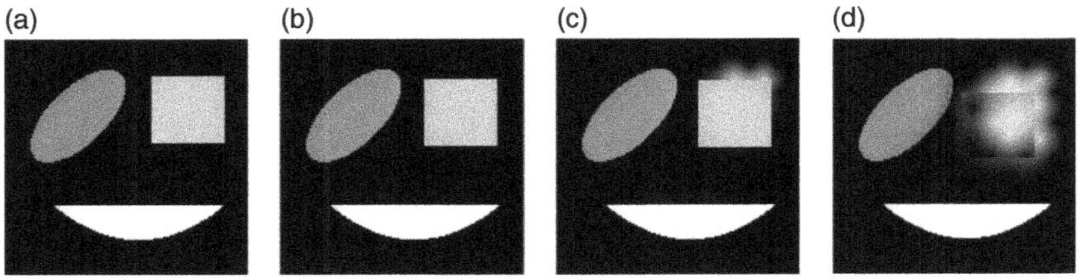

Figure 9.7 (a–d) Different reconstructions of f_G using the VSDK scheme with different scaling functions.

The resulting function f_G on Ω is given by $f_G = \chi_E + 1.5\chi_R + 2\chi_P$. The function f_G is shown in Figure 9.4.

To test the sensitivity of the VSDK scheme with respect to shifts of the scaling function ψ, we simulate applications where inaccurate scaling functions could occur. To this aim, we use the Matérn kernel $\varphi_{M,0}$, and we interpolate f_G and interpolate it at the Lissajous nodes $LS_2^{(10,11)}$ considering different scaling functions.

The results are displayed in Figure 9.7. From Figure 9.7a–d, we start with the correct scaling function (here, $\psi = f_G$) and progressively shift the rectangle in ψ toward the center. In Figure 9.8, we show the corresponding interpolation errors with respect to the original function f_G.

We observe that if the edges in the scaling function are correctly located or only slightly shifted, the reconstruction is accurate and no artifacts are visible. On the other hand, if ψ is more inaccurate, then the reconstruction is severely affected by distortions. Therefore, to effectively apply the VSDK scheme, it is necessary to correctly estimate the edges in the function to be reconstructed, as we will detail in Chapter 10.

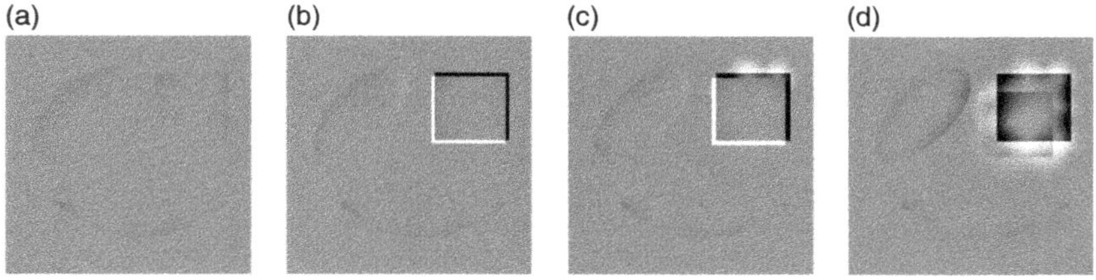

Figure 9.8 (a-d) Differences between the VSDK interpolants in Figure 9.7 and the original geometric phantom.

10
Application of VSDKs to Magnetic Particle Imaging

Magnetic particle imaging is an emerging tracer-based medical imaging technique measuring distributions of superparamagnetic nanoparticles using field-free point trajectories, often following Lissajous curves. The resulting tracer concentration frequently exhibits sharp edges between regions, making them ideal candidates for VSDK reconstruction. We present a shape-driven algorithm combining edge detection via SVM classification with VSDK interpolation. The approach is fully automatic, requiring only threshold selection for binary classification, and significantly outperforms standard polynomial schemes in both visual quality and quantitative accuracy metrics.

10.1 Extracting the Edges from the Data

The problem of edge detection in an image, often for segmentation purposes, has been extensively investigated in the literature because it arises in many applications. In this direction, several algorithms have been studied and widely applied (see, e.g., Solomon and Breckon, 2010, for a general overview). Many edge detection schemes require evaluating the function to be reconstructed on an underlying grid. For example, this is the case of the *Canny edge detector* (Canny, 1986), recently applied in De Marchi et al. (2017) in the context of MPI and of the *Sobel algorithm*, see Sharifi et al. (2002).

In our approximation setting, to apply such methods, we would need to compute a first reconstruction from the samples at our disposal. This is a notable drawback for two reasons.

(i) Calculating an additional reconstruction can be computationally costly.

(ii) As discussed in Section 9.2.5, standard interpolation algorithms can provide reconstructions that are affected by distortions and artifacts, which can cause possible problems in the subsequent edge detection process.

Therefore, it is preferable to look for image segmentation algorithms that can be applied directly to scattered data. A possible choice is to adopt machine learning classification algorithms. Among various classification methods, we take SVMs, which are discussed in detail below in Section 13.2.

Recalling the learning setting introduced in Problem 12.1.1 in Section 12.1, in our experiments the test instances are the points of the equispaced evaluation grid Ξ, while the training dataset consists of the set of Lissajous nodes LS_2^N defined in (9.5), which however needs to be labeled.

In general, the choice of labeling technique depends on the application. In Sections 10.1.1 and 10.1.2, we recall a couple of heuristic labeling strategies.

10.1.1 Using Thresholds on the Function Values

A simple method is to consider thresholds $a_0 < a_1 < \cdots < a_d$, $d \in \mathbb{N}$, for the definition of the labels. Indeed, if the function values $f(x_k), k = 1,\ldots,N$ are contained in the interval $[a_0, a_d)$, we can assign $\boldsymbol{x}_i$ to the class C_j if $a_{j-1} \leq f(\boldsymbol{x}_i) < a_j$, $j \in \{1,\ldots,d\}$.

10.1.2 Using Thresholds on RBF Interpolation Coefficients

As shown in Romani et al. (2018), variations in the expansion coefficients of an RBF interpolation can be used to detect the edges of a function. Therefore, thresholds on the absolute value of the RBF coefficients can be applied, as in the previous strategy. However, differently from the previous method, it is necessary to construct the associate RBF interpolant.

10.2 Background in Medical Imaging

For the following introduction, we refer mainly to Knopp and Buzug (2012). Over the last century, several tomographic imaging methods have been developed for medical applications. Every imaging technique is primarily characterized by the physical effects it accounts for to produce the image. Indeed, some methods measure variables directly linked to the tissue of interest, while others use a tracer injected into the body and subsequently detected to recover its spatial distribution.

Some well-known imaging techniques are computed tomography (CT), which computes the attenuation of X-rays; magnetic resonance imaging (MRI), which first adopts magnetic fields to excite the protons in the tissue and then measures their relaxation properties; positron-emission tomography (PET) and single-photon emission computed tomography (SPECT) that detect the position of a radioactive tracer. Some other characteristics of the cited methods are displayed in Table 10.1 (see Knopp and Buzug, 2012, table 1.1).

Table 10.1 Properties of some medical imaging methods

	CT	MRI	PET	SPECT
Spatial resolution	0.5 mm	1 mm	4 mm	10 mm
Temporal resolution	1 s	1 s–1 h	1 min	1 min
Sensitivity	Low	Low	High	High
Harmfulness	X-ray	Heating	β/γ radiation	γ radiation

While CT and MRI yield morphological information, tracer-based methods such as PET and SPECT provide information concerning functional activities; thus, they are fundamental, for example, in the detection of cancer cells. To obtain both morphological and functional details, combined methods such as PET/CT and PET/MRI have been developed.

10.3 Magnetic Particle Imaging

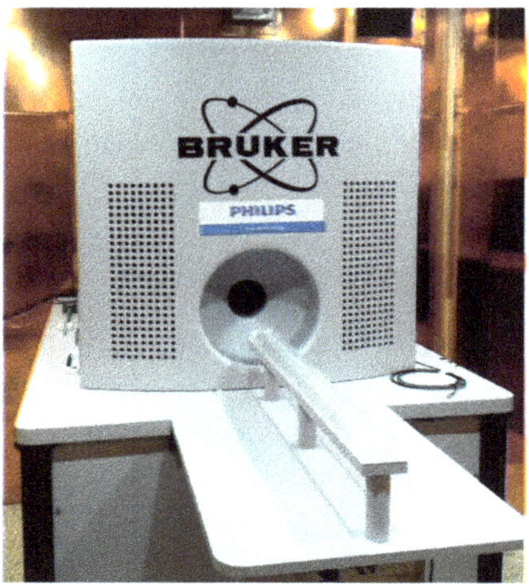

Figure 10.1 Bruker preclinical MPI scanner (from Open MPI data website (n.d.)).

In 2005, B. Gleich and J. Weizenecker (Gleich and Weizenecker, 2005) presented a new quantitative imaging method called MPI. It is a tracer-based technique in which the distribution of super paramagnetic iron oxide (SPIO) nanoparticles injected into the body is detected by superimposing multiple magnetic fields. Some remarkable features of MPI are the following (see Knopp and Buzug, 2012, chapter 1).

(i) It promises high spatial (<1 mm) and temporal (<0.1 s) resolution, thus it is applicable to real-time imaging.
(ii) Because of the usage of a tracer, it can be far more sensitive than MRI.
(iii) Unlike CT, PET, and SPECT, the MPI is free of ionizing radiation, since it uses just various magnetic fields.

Due to its resolution, MPI is well-suited for applications such as blood flow visualization, oncology imaging, and cell tracking (Wu et al., 2019; Yu et al., 2017). While MPI has already been proven effective in a preclinical setting (see Figure 10.1 for an example of an MPI preclinical scanner), the proof that it is clinically sensible is still pending due to technical challenges in upscaling to human-sized systems (Graeser et al., 2019).

Typically, in an MPI scanner, the signal acquisition proceeds as follows. A generated static nonuniform *gradient field* (also called *selection field*) contains a field free point (FFP), where the field magnitude is zero. As a consequence, the SPIO nanoparticle magnetization is saturated at all field-of-view positions, but in a small area around the FFP, where the nanoparticles are in the dynamic region of their magnetization curve. Then, a *drive field* is superimposed on the gradient field. The drive field moves the FFP in the field of view (see Figure 10.2). Moreover, the consequent change in the nanoparticles' magnetization induces a voltage that is measured by the receive coils. The spatial distribution of the tracer is therefore recovered by solving an ill-posed inverse problem, adopting proper regularization techniques (Knopp and Buzug, 2012, chapters 2 and 5).

Widely considered trajectories for the FFP are *Lissajous curves* (Knopp et al., 2009). Along such curves, the reconstruction can be performed by restricting to particular sampling points, as explained in Kaethner et al. (2016); these are the Lissajous nodes $LS_2^{\boldsymbol{N}}$ introduced in (9.5). The Lissajous nodes, and the corresponding polynomial reconstruction described in Erb et al. (2016), represent a theoretical extension of the theory concerning the Padua points (Bos et al., 2006, 2017).

When the nanoparticle density exhibits sharp edges, the Gibbs phenomenon occurs, affecting image reconstruction. In De Marchi et al. (2017), post-processing filters have been considered to mitigate the oscillations that result in the polynomial reconstruction scheme. Recently, in De Marchi et al. (2020a), the FNA introduced in Section 9.2.3 has also been applied in this

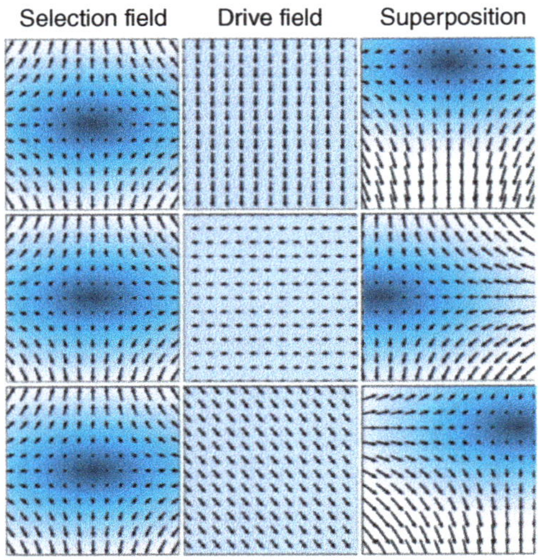

Figure 10.2 Selection and drive fields in MPI (from (Knopp and Buzug, 2012, figure 2.23)).

context. Here, we use the VSDK interpolation technique to effectively avoid ringing artifacts in an MPI application.

10.4 Results with VSDKs

Algorithm 10.4.1 (Shape-driven interpolation with VSDKs)

`INPUTS`: *Set of interpolation nodes*

$$X = \{\boldsymbol{x}_i,\ i = 1,\dots,N\} \subseteq \Omega,$$

a corresponding set of data values

$$F = \{f_i = f(\boldsymbol{x}_i),\ i = 1,\dots,N\},$$

and the desired evaluation point(s) $\boldsymbol{x} \in \Omega$.

`OUTPUTS`: *VSDK interpolant* $s^{\Psi}(\boldsymbol{x})$,

`Step 1`: *Extract the labels for X using F (see Section 10.1).*

`Step 2`: *Train the SVM algorithm with the points X and the obtained labels to construct the the scaling function* ψ *(see Section 10.1).*

`Step 3`: *Calculate the coefficient vector* $\boldsymbol{c}$ *of the VSDK interpolant.*

`Step 4`: *Evaluate the interpolant* $s^{\Psi}(\boldsymbol{x})$ *at* $\boldsymbol{x} \in \Omega$.

The dataset considered in De Marchi et al. (2020e) comprises MPI measurements reported in Kaethner et al. (2016) on a phantom consisting of three tubes filled with *Resovist*, a superparamagnetic iron oxide contrast agent. As described in Kaethner et al. (2016), we obtain a reconstruction of the nanoparticle density at the nodes $LS_2^{(33,32)}$, which is presented in Figure 10.3a. A related polynomial interpolant that exhibits ringing artifacts is shown in Figure 10.3b.

To construct the scaling function ψ and apply the VSDK scheme, we consider the Algorithm 10.4.1 as in De Marchi et al., 2020e. More precisely, for the labeling, we adopt the thresholding strategy described in Section 10.1 using $1/5$ of the maximal signal strength as a threshold for a binary classification. Then, we train the SVM algorithm taking the Gaussian kernel (see Section 13.2.2 for further details). The resulting scaling function is shown in Figure 10.3c.

Employing the Matérn kernel $\varphi_{M,0}$ for the VSDK interpolation process, we show the final interpolant for the given MPI data in Figure 10.3d. We observe that the VSDK scheme produces a reconstruction that is not distorted, unlike the polynomial case.

Moreover, in Figure 10.4 we provide another illustration (b) and (d) of Figure 10.3. More precisely, we present a diagonal view directed from the lower left to the upper right corner of the respective image. As in Figure 10.3, the mitigation of the Gibbs phenomenon is visible.

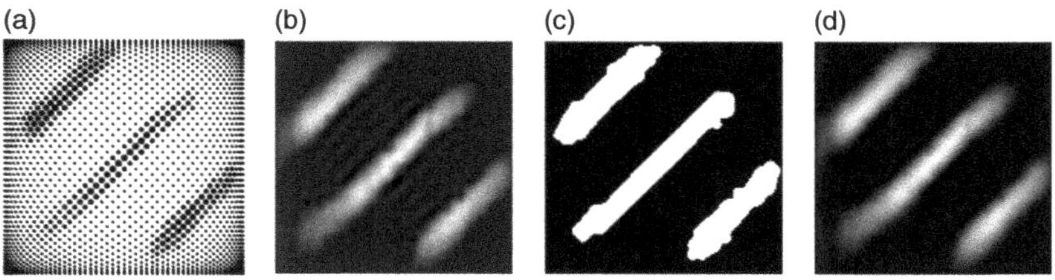

Figure 10.3 Comparison of different interpolation methods in MPI. The data at the Lissajous nodes $LS_2^{(33,32)}$ (a) are first interpolated using the polynomial scheme derived in Erb et al. (2016) (b). Using the constructed scaling function ψ (c), we obtain the VSDK interpolant by means of Algorithm 10.4.1 (d).

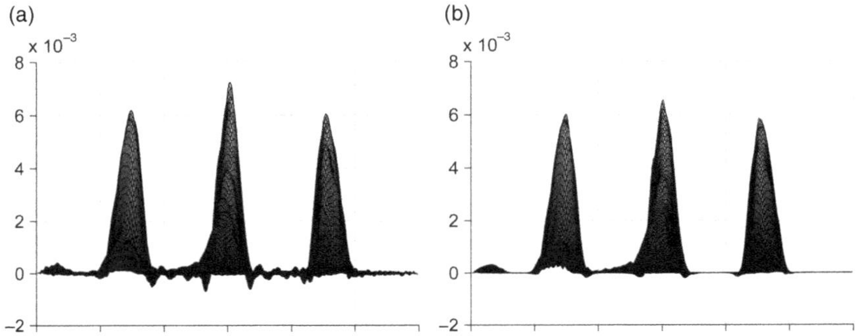

Figure 10.4 Comparison of polynomial (a) and VSDK interpolation (b) in MPI, considering a diagonal view from the lower left to the upper right corner.

11
Applications to Astronomical Imaging

As a second application of the VSKs, we address the problem of approximating parametric Fourier imaging problems via interpolation/extrapolation algorithms. It consists in interpolating via VSKs the scattered observations in the Fourier domain and then defining the sought approximation via the *projected Landweber iterative scheme*. We provide new theoretical insights, including error bounds in the image space and properties of the projected Landweber iterative scheme, both of which depend on the choice of the VSK scaling function. More precisely, because of the studies carried out in Chapter 16, we could use the VSKs for parametric Fourier imaging problems; indeed, the information encoded in an image reconstructed for a given parameter is transferred during the reconstruction process to a contiguous parameter value. Benchmark test cases in the field of astronomical imaging numerically demonstrate that the proposed scheme regularizes along the parameter direction, yielding reliable and interpretable results.

11.1 Background on Fourier Imaging Problems

Fourier imaging problems require the solution of the inverse Fourier transform problem when a limited number of noisy Fourier components of the imaging signal are at the disposal (Easton, 2010; Khare et al., 2015; Olson, 2017). This image reconstruction problem is ill-posed in the sense of Hadamard, and therefore the use of regularization methods is recommended to address its solution (Engl et al., 1996; Bertero and Piana, 2006). For specific but significant applications, the image formation model is generalized to encode the fact that the imaging device may record sets of Fourier samples for many different values of a specific parameter. Examples of this parametric Fourier imaging problem are dynamic MRI, where the parameter is time (Gamper et al., 2008; Sourbron and Buckley, 2013; Lingala et al., 2011; Ravishankar and Bresler, 2015), and imaging spectroscopy, where the parameter is energy Prévost et al. (2022); Zhang et al. (2022); Gary (2023); Piana et al. (2022).

As far as imaging spectroscopy is concerned, recent studies have already shown that interpolation–extrapolation schemes, in which an interpolation algorithm is applied in the spatial frequency domain and an out-of-band extrapolation approach is applied to the interpolated signal to obtain super-resolution effects, turns out to be effective (Larkman and Nunes, 2007; Tsao and Kozerke, 2012; Deng et al., 2012; Allavena et al., 2012; Perracchione et al., 2021, 2023). Our analysis provides the following:

(i) $\mathscr{L}_2$ error bounds in the image space that depend on the selection of the scaling function in the Fourier domain (see Chapter 16 too).
(ii) Bounds for the extrapolation iterative scheme that depend again on the selection of the scaling function.
(iii) New solutions for the definition of the VSK scaling function in the context of inverse problems that continuously depend on a parameter.

The theoretical findings have been validated by considering samples from the Reuven Ramaty high-energy solar spectroscopic imager (RHESSI) (Lin et al., 2002), which was a telescope recording X-rays from the Sun with the main purpose of observing solar flares (Piana et al., 2022). Within this framework, imaging spectroscopy consists of providing reconstructions of hard X-ray maps at different count energy channels.

11.2 The VSK Interpolation/Extrapolation Procedure

Given $g \in \mathscr{L}_2(\mathbb{R}^d) \cap \mathscr{C}(\mathbb{R}^d)$ and $f \in \mathscr{L}_2(\mathbb{R}^d)$ such that

$$(\mathrm{F}f)(\boldsymbol{\omega}) = g(\boldsymbol{\omega}), \quad \boldsymbol{\omega} \in \mathbb{R}^d, \tag{11.1}$$

where $\mathrm{F} : \mathscr{L}_2(\mathbb{R}^d) \to \mathscr{L}_2(\mathbb{R}^d)$ is the Fourier operator, we aim to recover f given a few samples of the function g. As the proposed scheme is based on interpolating in the Fourier domain and then extrapolating the resulting approximation, the function g must be continuous. Moreover, for simplicity, until Section 11.2.2 we limit our attention to nonparametric problems and assume that both f and g are continuous real-valued functions. Any extension to a complex function is straightforward by considering the real and imaginary parts separately.

11.2.1 Error Analysis for the Procedure

In practical scenarios, usually only a few measurements $\{g_i\}_{i=1}^n \in D \subseteq \mathbb{R}^d$, where D is a compact set, are available. This is the motivation for studying interpolation/extrapolation approaches. More formally, this means that the problem in (11.1) is indeed a discrete Fourier problem, where $\mathrm{F} = \mathrm{F} \cdot \chi_D$.

As only scattered measurements $\{g_i\}_{i=1}^n \in D \subseteq \mathbb{R}^d$ of the function g are available, one may think of interpolating via VSKs the given data to get a continuous function on the left-hand side of (11.1) as

$$s^{\Psi}(\boldsymbol{\omega}) := s(\boldsymbol{\omega}, \psi(\boldsymbol{\omega})) = \sum_{i=1}^{n} c_i \kappa((\boldsymbol{\omega}, \psi(\boldsymbol{\omega})), (\boldsymbol{\omega}, \psi(\boldsymbol{\omega}_i))), \quad \boldsymbol{\omega} \in \mathbb{R}^d.$$

Since the samples of g are contained in $D \subset \mathbb{R}^d$, and we lack information outside of this set, we extend s^{Ψ} to 0 in $\mathbb{R}^d \setminus D$. Specifically, first, we define:

$$\hat{s}^{\Psi}(\boldsymbol{\omega}) := \begin{cases} s^{\Psi}(\boldsymbol{\omega}), & \boldsymbol{\omega} \in D, \\ c, & \text{otherwise}, \end{cases}$$

being $c > 0$. Then we set $s^{\Psi}(\boldsymbol{\omega}) = \hat{s}^{\Psi}(\boldsymbol{\omega}) \cdot \rho(\boldsymbol{\omega})$, with

$$\rho(\boldsymbol{\omega}) = \begin{cases} 1, & \boldsymbol{\omega} \in D, \\ \rho_1(\boldsymbol{\omega}), & \boldsymbol{\omega} \in \mathbb{R}^d \setminus D. \end{cases}$$

Here, $\rho_1(\boldsymbol{\omega})$ is a smooth function that rapidly decreases to 0, so that the function ρ is continuous in $\mathbb{R}^d$.

Regarding ψ, since there are no issues with its continuity, we extend it by zero-padding to all of $\mathbb{R}^d$, that is,

$$\psi(\boldsymbol{\omega}) := \begin{cases} \psi(\boldsymbol{\omega}), & \boldsymbol{\omega} \in D, \\ 0, & \text{otherwise}. \end{cases}$$

Let $\bar{f}$ be the function reconstructed by taking the interpolating function on the left-hand side of (11.1), that is, $\bar{f}$ is so that

$$(\mathrm{F}\bar{f})(\boldsymbol{\omega}) = s^{\Psi}(\boldsymbol{\omega}).$$

Then we can show the following generic error estimate.

Proposition 11.2.1 *Assuming that the continuous functions $f, \bar{f}, g$, and s^{Ψ} belong to $\mathscr{L}_2(\mathbb{R}^d)$, the following stability estimate holds true*

$$\|f - \bar{f}\|_{\mathscr{L}_2(\mathbb{R}^d)} \leq \nu(\boldsymbol{\omega})\mu(D)^{1/2} + \|g - s^{\Psi}\|_{\mathscr{L}_2(\mathbb{R}^d \setminus D)},$$

where

$$\nu(\boldsymbol{\omega}) = \sup_{\boldsymbol{\omega} \in D} \left| g(\boldsymbol{\omega}) - s^{\Psi}(\boldsymbol{\omega}) \right|$$

Proof.

$$\begin{aligned}
\|f-\bar{f}\|_{\mathscr{L}_2(\mathbb{R}^d)} &= \|(\mathrm{F}^{-1}g)-(\mathrm{F}^{-1}s^{\Psi})\|_{\mathscr{L}_2(\mathbb{R}^d)} \\
&= \|g-s^{\Psi}\|_{\mathscr{L}_2(\mathbb{R}^d)} \le \|g-s^{\Psi}\|_{\mathscr{L}_2(D)} + \|g-s^{\Psi}\|_{\mathscr{L}_2(\mathbb{R}^d\setminus D)} \\
&\le \sup_{\boldsymbol{\omega}\in D}\left|g(\boldsymbol{\omega})-s^{\Psi}(\boldsymbol{\omega})\right|\mu(D)^{1/2} + \|g-s^{\Psi}\|_{\mathscr{L}_2(\mathbb{R}^d\setminus D)}.
\end{aligned}$$

□

Remark 11.2.1 *Considering Proposition 11.2.1, as in real applications we usually have information on a compact set of the frequency space, we are generally unable to bound* $\|g-s^{\Psi}\|_{\mathscr{L}_2(\mathbb{R}^d\setminus D)}$. *Nevertheless, it should be realistically negligible, as the function g should rapidly and smoothly decrease to zero.*

The first term in Proposition 11.2.1 can be bounded using the same arguments as in Proposition 16.1.1, and hence we immediately get

$$\left|\left(g-s^{\Psi}\right)(\boldsymbol{\omega})\right| \le \|g-s^{\Psi}_{\psi}\|_{\infty} + \|\Psi-\mathbf{g}\|_{\infty}\lambda^{\Psi}(\boldsymbol{\omega}). \tag{11.2}$$

Once more, the first term in (11.2) suggests that the scaling function ψ should be close to g on the whole domain.

11.2.2 Parametric Imaging Problems

For many kinds of signals, in physics we encounter pairs of functions related by an integral of the form

$$\tilde{g}(\boldsymbol{\omega},\beta) = \int_{\beta}^{+\infty} w(\boldsymbol{\omega},b)\eta(\beta,b)\,db, \tag{11.3}$$

for a fixed $\boldsymbol{\omega}\in\mathbb{R}^d$, and where $\beta\in\mathbb{R}$ is a given parameter that might depend on the hardware instrumentation. The reason why such integral transforms are common to many problems lies in the fact that (11.3) generalizes the quantum equivalence of the position and momentum representations of Schrödinger's equation (when $\boldsymbol{\omega}$ represents the space coordinate, and η is the Fourier kernel). In other words, w and $\tilde{g}$ are equivalent representations of the system under study. Hence, the general form of (11.3) can be seen as the common representation of signal formation in many fields, for example, interferometry, optics, radioastronomy, nuclear medicine, and inverse transport theory. For a fixed $\boldsymbol{\omega}\in\mathbb{R}^d$, we now assume that the real-valued functions $\tilde{g}, w$, and η are continuous and that the product between w and η is a Riemann-integrable function. To formally investigate how the scaling function may be selected for parametric imaging problems, we prove the following result.

Proposition 11.2.2 *Let η be a bounded and Lipschitz continuous function in β. Let η, w, and the product of the functions η and w be in $\mathscr{L}_2 \cap \mathscr{L}_1((\beta,+\infty))$. Let $\tilde{g}$ be the scaling function of the VSK interpolant, computed as in* (11.3). *Let β_1 and β_2 be larger than $\beta \in \mathbb{R}$. If $\beta_1 \longrightarrow \beta_2$ then $|\tilde{g}(\boldsymbol{\omega},\beta_1) - \tilde{g}(\boldsymbol{\omega},\beta_2)| \longrightarrow 0$.*

Proof. Without any restriction, we suppose $\beta_2 > \beta_1$. Then for a fixed $\boldsymbol{\omega} \in \mathbb{R}^d$, we have that

$$
\begin{aligned}
&|\tilde{g}(\boldsymbol{\omega},\beta_1) - \tilde{g}(\boldsymbol{\omega},\beta_2)| = \left| \int_{\beta_1}^{\infty} w(\boldsymbol{\omega},b)\eta(\beta_1,b)\,db - \int_{\beta_2}^{\infty} w(\boldsymbol{\omega},b)\eta(\beta_2,b)\,db \right| \\
&= \left| \int_{\beta_1}^{\beta_2} w(\boldsymbol{\omega},b)\eta(\beta_1,b)\,db + \int_{\beta_2}^{\infty} w(\boldsymbol{\omega},b)\eta(\beta_1,b)\,db \right. \\
&\left. - \int_{\beta_2}^{\infty} w(\boldsymbol{\omega},b)\eta(\beta_2,b)\,db \right| \\
&\le \int_{\beta_1}^{\beta_2} |w(\boldsymbol{\omega},b)\eta(\beta_1,b)|\,db + \int_{\beta_2}^{\infty} |w(\boldsymbol{\omega},b)(\eta(\beta_1,b) - \eta(\beta_2,b))|\,db \\
&\le |\beta_1 - \beta_2| \sup_{b\in[\beta_1,\beta_2]} (|\eta(\beta_1,b)|) \int_{\beta_1}^{\beta_2} |w(\boldsymbol{\omega},b)|\,db \\
&+ \sup_{b\in[\beta_2,\infty]} (|\eta(\beta_1,b) - \eta(\beta_2,b)|) \int_{\beta_2}^{\infty} |w(\boldsymbol{\omega},b)|\,db \\
&= |\beta_1 - \beta_2| \sup_{b\in[\beta_1,\beta_2]} (|\eta(\beta_1,b)|) \int_{\beta_1}^{\beta_2} |w(\boldsymbol{\omega},b)|\,db \\
&+ (|\eta(\beta_1,b^*) - \eta(\beta_2,b^*)|) \int_{\beta_2}^{\infty} |w(\boldsymbol{\omega},b)|\,db \\
&\le |\beta_1 - \beta_2| \sup_{b\in[\beta_2,\infty]} (|\eta(\beta_1,b)|) \int_{\beta_1}^{\beta_2} |w(\boldsymbol{\omega},b)|\,db + L|\beta_1 - \beta_2| \int_{\beta_2}^{\infty} |w(\boldsymbol{\omega},b)|\,db \\
&= |\beta_1 - \beta_2| \left(\sup_{[\beta_1,\beta_2]} (|\eta(\beta_1,b)|) \int_{\beta_1}^{\beta_2} |w(\boldsymbol{\omega},b)|\,db + L \int_{\beta_2}^{\infty} |w(\boldsymbol{\omega},b)|\,db \right).
\end{aligned}
$$

In this inequality chain, we have used the Hölder inequality and the fact that, denoting by T either the interval $[\beta_1,\beta_2] \subset [\beta,\infty]$ or $[\beta_2,\infty] \subset [\beta,\infty]$, for continuous and bounded functions we have

$$
||\eta(\beta_1,b)||_{L_\infty(T)} = \operatorname*{ess\,sup}_{b\in T} (|\eta(\beta_1,b)|) = \sup_{b\in T} (|\eta(\beta_1,b)|) < +\infty.
$$

Then, as w is in $\mathscr{L}_1((\beta,+\infty))$, that is, $||w||_{\mathscr{L}_1((\beta,+\infty))} < +\infty$, also $\int_{\beta_1}^{\beta_2} |w(\boldsymbol{\omega},b)|db$ and $\int_{\beta_2}^{\infty} |w(\boldsymbol{\omega},b)|\,db$ are both bounded. Hence, all quantities in the last step of the inequality chain are bounded and we can state that if $|\beta_1 - \beta_2| \longrightarrow 0$, then $|\tilde{g}(\boldsymbol{\omega},\beta_1) - \tilde{g}(\boldsymbol{\omega},\beta_2)| \longrightarrow 0$. □

This result offers the opportunity to define a scaling function in the VSKs setting which depends on the parameter β, that is, taking Ψ as in (11.3). Hence, supposing to have some information on the sought images in the frequency plane for a given parameter β_2, we can use it to define the scaling function for reconstructing the target at a sufficiently close parameter β_1, thus obtaining smoothing constraints in successive reconstructions. In view of this, we now need to investigate how the β-dependent scaling function affects the inversion step, which is performed via the so-called Landweber iterative scheme.

11.2.3 Analysis of the Landweber Scheme

Because in applications we have scattered data only in a compact domain, once the interpolation step has been performed, simply applying the inverse Fourier transform to the interpolated data would lead to significant drawbacks, like as ringing effects in the recovered image (Lagendijk et al., 1988). A possible way to overcome such limitations is to use a constrained inversion technique. More precisely, if we denote by $s^{\Psi}\chi_D$ the interpolant on D of the scattered observations, then

$$\bar{f}_{\text{nb}}(\boldsymbol{x}) = \left(\mathrm{F}^{-1}(s^{\Psi}\chi_D)\right)(\boldsymbol{x}),$$

is the noisy band-limited approximation of the function f.

For functions $f \in C_+$, where $C_+ := \{f \in \mathscr{L}_2(\mathbb{R}^d) \ : \ f \geq 0 \text{ a.e}\}$ denotes the closed convex cone of nonnegative functions in $\mathscr{L}_2(\mathbb{R}^d)$, a stable approximation can be computed via the projected Landweber iteration according to the following steps:

1. Initialize the iterative scheme as $f^{(0)} = 0$.
2. For a given threshold $\tau \in (0,2)$ and starting from $f^{(k)}$, compute

   $$\mathrm{F}(f^{(k+1)}) = \tau \mathrm{F}(\bar{f}_{\text{nb}}) + (1 - \tau\chi_D)\mathrm{F}(\tilde{f}^{(k)}), \quad k = 1,2,\ldots.$$

   and apply the nonnegativity constraint by means of the projection

   $$\tilde{f}^{(k+1)} = \mathrm{P}_+ f^{(k+1)},$$

   where

   $$\left(\mathrm{P}_+ f^{(k+1)}\right)(\boldsymbol{x}) = \begin{cases} 0, & \text{if } \Re(f^{(k+1)}(\boldsymbol{x})) < 0, \\ f^{(k+1)}(\boldsymbol{x}), & \text{otherwise.} \end{cases}$$

3. Continue with step 2, until

   $$\|\tilde{f}^{(k+1)} - \tilde{f}^{(k)}\|_{\mathscr{L}_2(\mathbb{R}^d)} \leq \delta,$$

where δ is a given tolerance. Therefore, we consider $\tilde{f} := \tilde{f}^{(k+1)}$ as the approximated solution of f.

Remark 11.2.2 *We highlight that the procedure is a* modified *Landweber method, since it is not a fixed-point method but depends on the initial value $\bar{f}_{nb}$ (from which the convergence depends). In addition, due to additional projections, iterations may lose the band-limited property even starting from an initially band-limited function. Finally, since operationally we do not have information outside the domain D, unlike the standard Landweber method, which guarantees convergence throughout $\mathbb{R}^d$, this method can guarantee convergence toward the exact solution only in the spectral domain D.*

To investigate the influence of the β-dependent definition of the VSK scaling function, a crucial point is to determine how the final approximated solutions are linked to the interpolants.

Theorem 11.2.1 *Let D a compact set in $\mathbb{R}^d$, and $\mathscr{P}_-$ be defined as*

$$\left(\mathrm{P}_- f^{(k+1)}\right)(\boldsymbol{x}) = \begin{cases} f^{(k+1)}(\boldsymbol{x}), & \text{if } \Re(f^{(k+1)}(\boldsymbol{x})) < 0, \\ 0, & \text{otherwise}. \end{cases}$$

For a fixed β, and a given tolerance $\delta > 0$, then

$$\left| \|\mathrm{F}\bar{f}_{nb} - \mathrm{F}\tilde{f}^{(k)}\|_{\mathscr{L}_2(D)} - \|\mathrm{FP}_-\mathrm{f}^{(k+1)}\|_{\mathscr{L}_2(\mathbb{R}^d)} \right| \leq \delta.$$

Proof. Let us suppose that $\|\tilde{f}^{(k+1)} - \tilde{f}^{(k)}\|_{\mathscr{L}_2(\mathbb{R}^d)} \leq \delta$, and denoting by $\tilde{\tilde{f}} := \mathrm{P}_-\mathrm{f}$, we get that

$$\begin{aligned}
\delta &\geq \|\tilde{f}^{(k+1)} - \tilde{f}^{(k)}\|_{\mathscr{L}_2(\mathbb{R}^d)} = \|\mathrm{F}f^{(k+1)} - \mathrm{F}\tilde{f}^{(k)} - \mathrm{F}\tilde{\tilde{f}}^{(k+1)}\|_{\mathscr{L}_2(\mathbb{R}^d)} \\
&= \|\tau\mathrm{F}\bar{f}_{nb} - \tau\chi_D\mathrm{F}\tilde{f}^{(k)} - \mathrm{F}\tilde{\tilde{f}}^{(k+1)}\|_{\mathscr{L}_2(\mathbb{R}^d)} \\
&\geq \left|\tau\|\mathrm{F}\bar{f}_{nb} - \mathrm{F}\tilde{f}^{(k)}\|_{\mathscr{L}_2(D)} - \|\mathrm{F}\tilde{\tilde{f}}^{(k+1)}\|_{\mathscr{L}_2(\mathbb{R}^d)}\right|.
\end{aligned}$$

□

We are now interested in understanding how the maps recovered, considering contiguous parameters are linked to their reconstruction via Landweber.

Proposition 11.2.3 *Given two real parameters $\beta_1 < \beta_2$, we have that*

$$\begin{aligned}
\|\tilde{f}_{\beta_2} - \tilde{f}_{\beta_1}\|_{\mathscr{L}_2(\mathbb{R}^d)} \leq& \|\mathrm{F}f_{\beta_2}^{(k+1)} - \mathrm{F}f_{\beta_1}^{(k+1)}\|_{\mathscr{L}_2(\mathbb{R}^d \setminus D)} + \\
&\left(1 - (1-\tau)^{k+1}\right)\|\mathrm{F}\bar{f}_{nb,\beta_2} - \mathrm{F}\bar{f}_{nb,\beta_1}\|_{\mathscr{L}_2(D)} \\
&+ \sum_{l=0}^{k-1}(1-\tau)^{\ell+1}\|\mathrm{F}\tilde{\tilde{f}}_{\beta_2}^{(k-l)} - \mathrm{F}\tilde{\tilde{f}}_{\beta_1}^{(k-l)}\|_{\mathscr{L}_2(D)}.
\end{aligned}$$

Proof. By the iterative scheme, we have

$$\|\tilde{f}_{\beta_2} - \tilde{f}_{\beta_1}\|_{\mathscr{L}_2(\mathbb{R}^d)} = \|\tilde{f}_{\beta_2}^{(k+1)} - \tilde{f}_{\beta_1}^{(k+1)}\|_{\mathscr{L}_2(\mathbb{R}^d)} \leq \|f_{\beta_2}^{(k+1)} - f_{\beta_1}^{(k+1)}\|_{\mathscr{L}_2(\mathbb{R}^d)}$$
$$= \|\mathrm{F}f_{\beta_2}^{(k+1)} - \mathrm{F}f_{\beta_1}^{(k+1)}\|_{\mathscr{L}_2(\mathbb{R}^d)}$$
$$\leq \|\mathrm{F}f_{\beta_2}^{(k+1)} - \mathrm{F}f_{\beta_1}^{(k+1)}\|_{\mathscr{L}_2(\mathbb{R}^d\setminus D)} + \|\mathrm{F}f_{\beta_2}^{(k+1)} - \mathrm{F}f_{\beta_1}^{(k+1)}\|_{\mathscr{L}_2(D)}$$
$$\leq \|\mathrm{F}f_{\beta_2}^{(k+1)} - \mathrm{F}f_{\beta_1}^{(k+1)}\|_{\mathscr{L}_2(\mathbb{R}^d\setminus D)} + \tau\|\mathrm{F}\bar{f}_{nb,\beta_2} - \mathrm{F}\bar{f}_{nb,\beta_1}\|_{\mathscr{L}_2(D)}$$
$$+ (1-\tau)\|\mathrm{F}\tilde{f}_{\beta_2}^{(k)} - \mathrm{F}\tilde{f}_{\beta_1}^{(k)}\|_{\mathscr{L}_2(D)}$$
$$\leq \|\mathrm{F}f_{\beta_2}^{(k+1)} - \mathrm{F}f_{\beta_1}^{(k+1)}\|_{\mathscr{L}_2(\mathbb{R}^d\setminus D)} + \tau\|\mathrm{F}\bar{f}_{nb,\beta_2} - \mathrm{F}\bar{f}_{nb,\beta_1}\|_{\mathscr{L}_2(D)}$$
$$+ (1-\tau)\|\mathrm{F}f_{\beta_2}^{(k)} - \mathrm{F}f_{\beta_1}^{(k)}\|_{\mathscr{L}_2(D)} + (1-\tau)\|\mathrm{F}\tilde{\tilde{f}}_{\beta_2}^{(k)} - \mathrm{F}\tilde{\tilde{f}}_{\beta_1}^{(k)}\|_{\mathscr{L}_2(D)}$$
$$= \|\mathrm{F}f_{\beta_2}^{(k+1)} - \mathrm{F}f_{\beta_1}^{(k+1)}\|_{\mathscr{L}_2(\mathbb{R}^d\setminus D)} + \tau\|\mathrm{F}\bar{f}_{nb,\beta_2} - \mathrm{F}\bar{f}_{nb,\beta_1}\|_{\mathscr{L}_2(D)}$$
$$+ (1-\tau)\tau\|\mathrm{F}\bar{f}_{nb,\beta_2} - \mathrm{F}\bar{f}_{nb,\beta_1}\|_{\mathscr{L}_2(D)} + (1-\tau)\|\mathrm{F}\tilde{\tilde{f}}_{\beta_2}^{(k)} - \mathrm{F}\tilde{\tilde{f}}_{\beta_1}^{(k)}\|_{\mathscr{L}_2(D)}$$
$$+ (1-\tau)^2\|\mathrm{F}\tilde{f}_{\beta_2}^{(k-1)} - \mathrm{F}\tilde{f}_{\beta_1}^{(k-1)}\|_{\mathscr{L}_2(D)}$$
$$\leq \cdots \leq \|\mathrm{F}f_{\beta_2}^{(k+1)} - \mathrm{F}f_{\beta_1}^{(k+1)}\|_{\mathscr{L}_2(\mathbb{R}^d\setminus D)} +$$
$$\left(\tau\sum_{l=0}^{k}(1-\tau)^l\right)\|\mathrm{F}\bar{f}_{nb,\beta_2} - \mathrm{F}\bar{f}_{nb,\beta_1}\|_{\mathscr{L}_2(D)}$$
$$+ \sum_{l=0}^{k-1}(1-\tau)^{\ell+1}\|\mathrm{F}\tilde{\tilde{f}}_{\beta_2}^{(k-l)} - \mathrm{F}\tilde{\tilde{f}}_{\beta_1}^{(k-l)}\|_{\mathscr{L}_2(D)}$$
$$= \|\mathrm{F}f_{\beta_2}^{(k+1)} - \mathrm{F}f_{\beta_1}^{(k+1)}\|_{\mathscr{L}_2(\mathbb{R}^d\setminus D)} +$$
$$\left(1-(1-\tau)^{k+1}\right)\|\mathrm{F}\bar{f}_{nb,\beta_2} - \mathrm{F}\bar{f}_{nb,\beta_1}\|_{\mathscr{L}_2(D)}$$
$$+ \sum_{l=0}^{k-1}(1-\tau)^{\ell+1}\|\mathrm{F}\tilde{\tilde{f}}_{\beta_2}^{(k-l)} - \mathrm{F}\tilde{\tilde{f}}_{\beta_1}^{(k-l)}\|_{\mathscr{L}_2(D)}.$$

□

Proposition 11.2.3 states that we can bound the difference between solutions at different parameter values by the $\mathscr{L}_2$ norm of the interpolants in D. Specifically, if we use a scaling function defined by contiguous values of β, we are essentially approximating by imposing constraints, that is, ensuring a lower variation norm.

Considering Proposition 11.2.3, as stated in Remark 11.2.1, in real applications we usually have no information outside a compact Fourier domain, so we reasonably suppose that $\|\mathrm{F}f_{\beta_2}^{(k+1)} - \mathrm{F}f_{\beta_1}^{(k+1)}\|_{\mathscr{L}_2(\mathbb{R}^d\setminus D)}$ is *small*. Nevertheless, we can bound the second and third terms within the compact Fourier domain.

As far as the last term is concerned, we again expect that its contribution is negligible outside D.

11.3 Applications to Astronomical Imaging

In this section, we focus on solar hard X-ray imaging and, specifically, we test our interpolation/extrapolation procedure in the framework of the NASA RHESSI mission (Lin et al., 2002).

From now on, we will denote the function g by V and f by I, following the notation typically used in this astronomical imaging scenario. The main purpose of the RHESSI system is to approximate the image of the unknown source flux emitted during a solar flare, given its Fourier transform, namely V, that is, the imaging problem reads as:

$$V(\boldsymbol{\omega}) = (\mathrm{F}I)(\boldsymbol{\omega}).$$

More precisely, the data provided by RHESSI consist of a set of scattered observations, called visibilities, associated with experimental measurements of the Fourier transform of the incoming photon flux, sampled at specific points (associated to the hardware detectors) in the spatial frequency plane lying on circles characterized by increasing radii (see Figure 11.1).

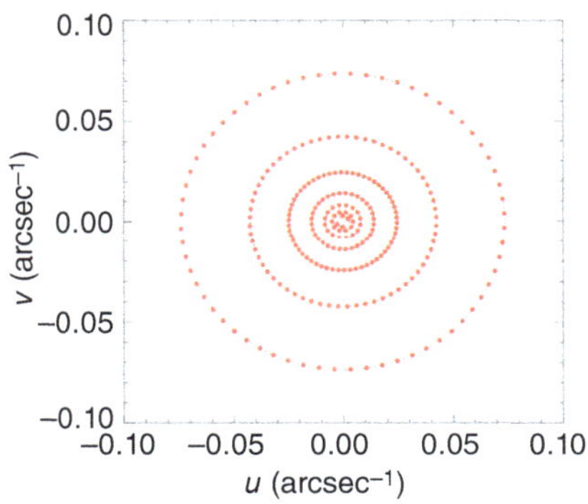

Figure 11.1 Frequency points sampled by the RHESSI instrument. The sampled points lie on seven circles (only detectors 3–9 are used), with increasing radii. The number of samples is about 250; it depends on the signal-to-noise ratio and changes from flare to flare. Here $\boldsymbol{\omega} = (u, v)$.

The RHESSI visibilities are measured at specific energy channels; that is, these energy ranges play the role of β in (11.3). Within this particular context, (11.3) relates the photon and electron visibilities, indeed as shown in Piana et al.

(2007), for a fixed $\boldsymbol{\omega}$, the link between $V(\boldsymbol{\omega};\ell)$ and $w(\boldsymbol{\omega};e)$, which respectively are the photon and electron visibilities, is given by

$$V(\boldsymbol{\omega};\ell) = \frac{1}{4\pi R^2}\int_{e=\ell}^{\infty} w(\boldsymbol{\omega};e)\eta(\ell,e)\,de\,,$$

where R is the distance from the source to the instrument, ℓ and e are photon and electron energy, respectively, and $\eta(\ell,e)$ is the bremsstrahlung cross-section, which gives the probability that a photon whose energy is ℓ is emitted by an electron whose energy is $e \geq \ell$. Considering the nature of the signal we are working with, that is, photon counts, the solution map is always nonnegative, and this implies that the positivity constraints in the Landweber iterations work in this setting too.

Corollary 11.3.1 *Under the same assumptions of Proposition 11.2.2, and fixing* $e > \ell_2 > \ell_1$, *if* $\ell_1 \longrightarrow \ell_2$ *then* $|V(\boldsymbol{\omega},\ell_1) - V(\boldsymbol{\omega},\ell_2)| \longrightarrow 0$.

Proof. Following the same steps of Proposition 11.2.2, we get:

$$|V(\boldsymbol{\omega},\ell_1) - V(\boldsymbol{\omega},\ell_2)| \leq \frac{1}{4\pi R^2}\left(\int_{e=\ell_1}^{\ell_2} |w(\boldsymbol{\omega},e)\eta(\ell_1,e)|\,de \right. \\ \left. + \int_{e=\ell_2}^{\infty} |w(\boldsymbol{\omega},e)||\eta(\ell_1,e) - \eta(\ell_2,e)|\,de\right).$$

Then, as shown in Massone et al. (2008), if $\ell_1 \longrightarrow \ell_2$, then $\int_{\ell_1}^{\ell_2} |w(\boldsymbol{\omega},e)\eta(\ell_1,e)|\,de \longrightarrow 0$, and the thesis follows. □

This suggests using as a scaling function dependent on the energy range. Specifically, let us suppose that we have a reconstructed image $\tilde{I}_{\ell_2} := \tilde{I}_{\ell_2}(\cdot,\ell_2)$ at the energy range ℓ_2. Then, our interpolant at the sought energy range ℓ_1 will be constructed with VSKs, and as scaling function $\tilde{V}_{\ell_2} := \tilde{V}(\cdot,\ell_2)$ we take:

$$\tilde{V}_{\ell_2}(\boldsymbol{\omega}) = (\mathrm{F}\tilde{I}_{\ell_2})(\boldsymbol{\omega}).$$

Then, we recover $\tilde{I}_{\ell_1}$ by approximating via the projected Landweber iterative scheme the following inverse problem:

$$s^{\tilde{V}_{\ell_2}}(\boldsymbol{\omega}) = (\mathrm{F}\tilde{I}_{\ell_1})(\boldsymbol{\omega}).$$

11.4 Results with Real Data

We test our procedure on a flare observed by RHESSI on July 03, 2002, considering the time window between 02:10:13 and 02:11:13 UT. RHESSI spectral resolution was $1-2$ keV below 1 MeV (Piana et al., 2022); therefore, we extend the discussion in the previous sections regarding the β parameter, which is connected to the hardware parameter, to sequences of contiguous energy bins. Specifically, we have considered two 3-keV-wide channels, starting from 25 to 28 keV, and nine 2-keV-wide channels, from $20-22$ keV to $4-6$ keV. For this event, RHESSI provided $N = 248$ visibilities at each energy channel. Figure 11.2 shows the reconstruction of the event in the considered energy intervals, using the interpolation/extrapolation procedure. The reconstruction used to trigger the scheme is the one provided by the maximum entropy method (MEM_GE) (Massa et al., 2020) in the energy range $25-28$ keV. From the energy range $22-25$ keV, each map is obtained by using as a scaling function in the interpolation step the discrete Fourier transform of the map retrieved in the previous (higher) energy channel. Therefore, given a succession of energy channels $\{\ell_l\}_{l=1}^{L}$, where $\ell_{l+1} < \ell_l$, we consider a sequence of scaling functions, $\{\tilde{V}_{\ell_l}\}_{l=1}^{L}$, and interpolants, to iteratively approximate $\tilde{I}_{\ell_{l+1}}$ via the relation:

$$s^{\tilde{V}_{\ell_l}}(\boldsymbol{\omega}) = (\mathrm{F}\tilde{I}_{\ell_{l+1}})(\boldsymbol{\omega})$$

where

$$\tilde{V}_{\ell_l}(\boldsymbol{\omega}) = (\mathrm{F}\tilde{I}_{\ell_l})(\boldsymbol{\omega}), \quad l = 1, \ldots, L.$$

We refer the reader to Volpara et al. (2025) for further details.

An advantage of this approach is that the visibilities in continuous energy channels are correlated. Thus, as shown in Figure 11.3, regularized photon spectra can be obtained: local count spectra derived from the regularized maps shown in Figure 11.2 are compared with those obtained from maps reconstructed using MEM_GE, when applied independently at each energy channel. The interpolation/extrapolation procedure ensures numerical stability. It reduces uncertainties (obtained by a confidence strip approach), as evidenced by the black error bars, which are consistently smaller than those associated with the unregularized (uncorrelated) spectra (in pink).

Finally, the theoretical results given in Section 11.2.3 are confirmed by Figures 11.4 and 11.5, obtained by setting the relaxation parameter for Landweber iteration $\tau = 0.2$. Specifically, each panel of Figure 11.4 corresponds to a specific energy channel. For each iteration, the red line shows the global distance between two subsequent steps of the Landweber iterative scheme. In contrast, the black line shows the global distance between the maps at each iteration of the Landweber process and the initial interpolants, taking into account the contribution related to $\mathrm{P\_I}^{(k+1)}$, as defined in the left-hand side of Theorem 11.2.1.

Figure 11.5 shows the result established in Proposition 11.2.3. It is important to note that no information is available outside the compact Fourier domain

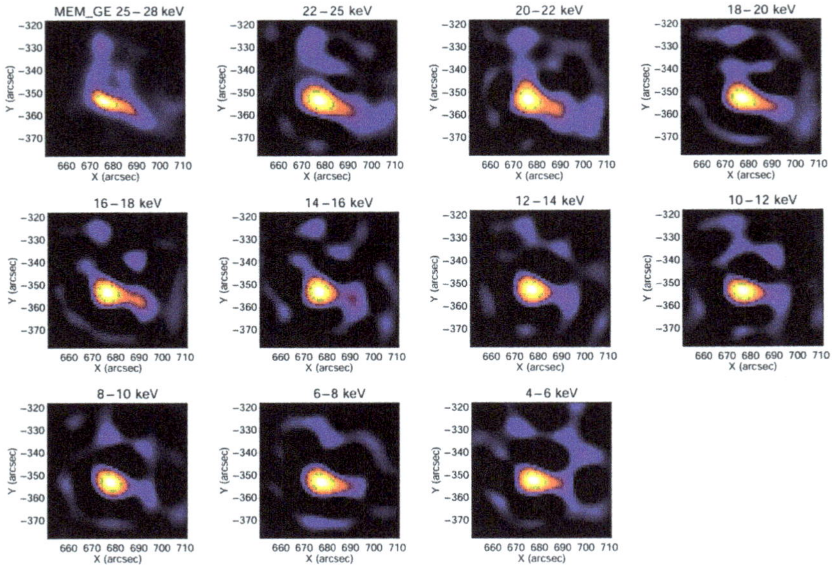

Figure 11.2 Reconstructions of the July 03, 2002, event, in the time range 02:10:13–02:11:13 UT for the energy intervals shown, obtained by the interpolation/extrapolation iterative method, using the MEM_GE reconstruction at the first panel as triggering image.

D, and thus, the first term on the right-hand side is not taken into account. Additionally, the same figure demonstrates that the negative contribution in the Landweber scheme (associated with $\mathrm{P\_I}^{(\mathrm{k}+1)} = \tilde{\tilde{\mathrm{I}}}^{(\mathrm{k}+1)}$) is negligible compared to the map obtained after applying the positivity constraint (i.e., $\tilde{I}^{(k+1)}$), blue line in figure.

These insights collected in this chapter could guide the development of more refined VSK-based algorithms for sparser imaging systems, especially given the growing interest in nano-satellites that provide only a few observations. Overall, this work lays the groundwork for further exploration of VSKs across a range of inverse problem settings, including medical imaging.

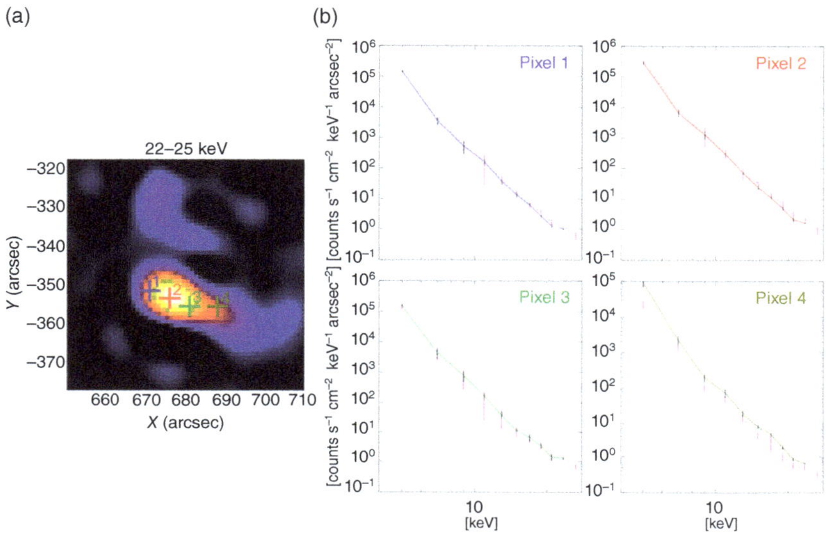

Figure 11.3 (a) Selected pixels are indicated with colored crosses. (b) Corresponding pixel-wise spectra obtained from photon images, in the case of iterative interpolation/extrapolation procedure (solid line) and MEM_GE (dashed line). The pixels selected in (a) and their respective spectra are indicated with the same color.

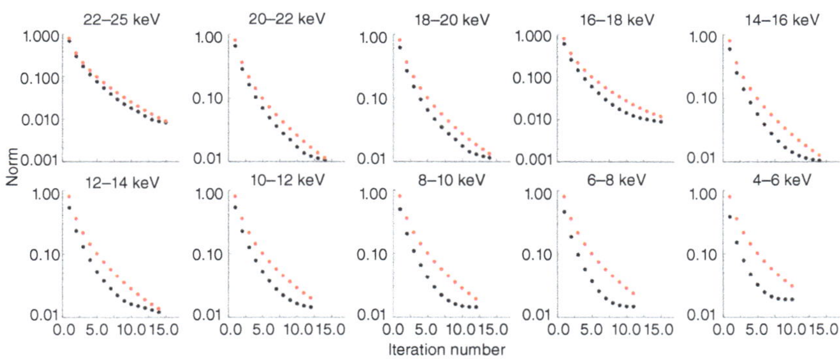

Figure 11.4 Numerical results for Theorem 11.2.1 in the case of RHESSI data. Red lines (upper) show the relative 2-norm of the pixel-wise distance between maps obtained from two consecutive iterations of projected Landweber scheme. Black lines (lower) show the relative 2-norm of the distance between the maps at each iteration of the Landweber process and the initial interpolants.

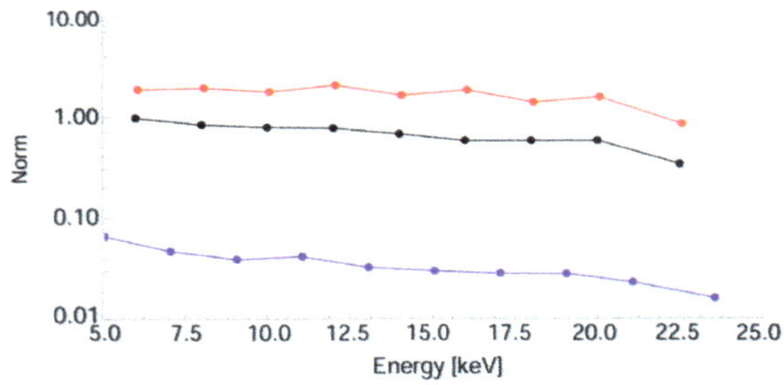

Figure 11.5 Numerical results for the Proposition 11.2.3 in the case of RHESSI data. Blue lines (bottom) show the differences between final reconstructions obtained at two nearby energy parameters as a function of iteration and as a function of energy step. The figure contains separate plotted contributions corresponding to terms appearing in the inequality of Proposition 11.2.3: a term for differences outside the compact Fourier domain (black lines, middle) and a term proportional to the difference between noisy band-limited interpolants on the Fourier compact set.

12
Kernels in Machine Learning

Kernel methods provide a unifying framework that connects approximation theory and statistical learning, enabling nonlinear pattern recognition in high-dimensional feature spaces without explicitly computing feature maps. We discuss the Vapnik–Chervonenkis dimension and Rademacher complexity (RC) as capacity measures, and present error bounds that balance empirical risk (training error) with model complexity. The bias-variance trade-off and regularization principles are discussed in the context of avoiding overfitting. Properties of positive definite kernels, including closure under addition, multiplication, and composition, enable flexible kernel design.

12.1 Learning from Data

We present supervised learning problem (Problem 12.1.1) focusing on the binary case. However, we note that Problem 12.1.1 can be easily generalized to the multiclass framework.

Problem 12.1.1 (Binary supervised learning) *Let* $\Omega \subset \mathbb{R}^d$ *and let* $X = \{\boldsymbol{x}_k, k = 1,\dots,N\} \subset \Omega$ *be a set of data (or inputs), where* $d,N \in \mathbb{N}$ *and* $d,N \geq 1$, *and let* $Y = \{+1,-1\}$. *Let us suppose that each element in X belongs to one out of two classes (also called labels or outputs), which are formally denoted with* $+1$ *and* -1. *Thus, we can consider the labeled dataset*

$$Z := \{(\boldsymbol{x}_k, y_k),\ \boldsymbol{x}_k \in X,\ y_k \in Y\} \subset X \times Y,$$

where the elements in X are linked to the respective class. We also refer to Z as the training dataset.

Given Z, we consider the binary supervised learning problem of finding a function $s : \Omega \longrightarrow Y$ *such that:*

(i) it models the input–output relation in Z;
(ii) it models the input–output relation of unseen labeled test instances $\{(\boldsymbol{\xi}_k, y_k)\}_{k=1,\ldots,M} \subset (\Omega \setminus X) \times Y$*, where* $M \geq 1$.

As we shall discuss, it is easy to select a function s that satisfies requirement (i) in Problem 12.1.1, but the usage of such a function in the classification of unseen test data, according to requirement (ii), could be unsatisfactory or even meaningless. In fact, the *generalization* capability required in (ii) is the crucial issue of the learning problem, which is also denoted as the pattern recognition problem (Schölkopf and Smola, 2002, Section 1.3).

Remark 12.1.1 *For regression tasks, Problem 12.1.1 can be reformulated by letting* $Y \subset \mathbb{R}$*. Although we are mainly interested in classification problems, in Section 13.1 we present a kernel-based regression model.*

A concrete strategy for finding a solution to Problem 12.1.1 will be presented in Section 13.2, after the introduction of the necessary ingredients in Section 12.2.
In this section, we examine Problem 12.1.1 from the point of view of statistical learning theory, mainly referring to Schölkopf and Smola (2002, section 1.3 and chapter 5) and Shawe-Taylor and Cristianini (2004, chapter 4). The books by Vapnik (1998, 1995, 1982) are also important references. As will be clear, we take a *frequentist* approach. For an overview of *Bayesian* approaches, we refer, for example, to Murphy (2012).

12.1.1 Statistical Learning Theory and VC Dimension

Let us suppose that both training and test data are sampled independently from some fixed unknown probability distribution $p(\boldsymbol{x}, y)$ on $\Omega \times Y$. Given a classifying function $s : \Omega \longrightarrow Y$, to measure the correctness of the resulting classification we consider the *zero-one loss function*

$$c(\boldsymbol{x}, y, s) := \frac{1}{2}|s(\boldsymbol{x}) - y|,$$

which is 0 if $s(\boldsymbol{x}) = y$ and 1 otherwise.
In this framework, considering the information in the training dataset Z and the properties (i) and (ii) required in Problem 12.1.1, we take the so-called *empirical risk*

$$\hat{e}(Z, s) := \frac{1}{N} \sum_{k=1}^{N} c(\boldsymbol{x}_k, y_k, s), \tag{12.1}$$

which is the empirical mean value of c fixed Z, and the so-called *generalization risk*

$$e(s) := \int_{\Omega \times Y} c(\boldsymbol{x}, y, s)\, \mathrm{d}p(\boldsymbol{x}, y), \tag{12.2}$$

which is the mean value of c averaged over all possible test samples generated from $p(\boldsymbol{x}, y)$. We point out that both generalization and empirical risk can be defined considering other loss functions, provided that the integral in (12.2) exists. In our setting, the generalization risk represents the average rate or probability of misclassification.

As previously noted, minimizing the empirical risk often does not necessarily correspond to a low generalization risk. Indeed, the following example suggests that the generalization risk can not be bounded by the empirical risk alone.

Example 12.1.1 *The function $g : \Omega \longrightarrow Y$ defined as*

$$g(\boldsymbol{x}) := \begin{cases} y_k, & \text{if } \boldsymbol{x} = \boldsymbol{x}_k, \\ 1, & \text{otherwise}, \end{cases}$$

provides $\hat{e}(Z, g) = 0$, but it seems to be a trivial and poor choice in terms of generalization capability.

Instead of discussing all possible classifiers on Ω individually, it is reasonable to restrict ourselves to a class of functions $\mathscr{S} \subset \{g \,|\, g : \Omega \longrightarrow Y\}$, which is defined according to certain specific properties or some geometric intuitions, and take $s \in \mathscr{S}$ to represent the entire class. For example, if $\boldsymbol{w} \cdot \boldsymbol{x} + b = 0$ is a hyperplane in $\mathbb{R}^d$, $\mathscr{S}$ could be defined as the set of functions $s(\boldsymbol{x}) = \mathrm{sgn}(\boldsymbol{w} \cdot \boldsymbol{x} + b)$ varying the parameters $\boldsymbol{w} \in \mathbb{R}^d$ and $b \in \mathbb{R}$; in fact, this is the setting considered in the classification with SVMs (see Section 13.2).

Statistical learning theory proves that it is important to restrict to a class $\mathscr{S}$ whose *capacity* is suitable for the amount of training data available (see Schölkopf and Smola (2002, chapter 5) or Vapnik (1998, chapters 5 and 6)). In principle, the capacity of a class of classifying functions is its capability in expressing complex patterns. This general idea can be formalized in different ways, of which the *Vapnik–Chervonenkis* (VC) dimension is well known (Vapnik and Chervonenkis, 1971).

Definition 12.1.1 Let $X = \{\boldsymbol{x}_k,\ k = 1, \ldots, N\}$ be a set of N data and let $Y = \{+1, -1\}$ be the set of labels. Let $Z_1, \ldots, Z_{2^N}$ be all the different datasets obtainable by taking all possible configurations of labels assigned to the data. A class $\mathscr{S}$ shatters the set X if for every dataset Z_i, $i = 1, \ldots, 2^N$, there exists a function $s : \Omega \longrightarrow Y$, $s \in \mathscr{S}$, such that $\hat{e}(Z, s) = 0$.

Definition 12.1.2 The VC dimension of a class $\mathscr{S}$ of classifying functions is the largest natural number N such that there exists a set X of N points that can be shattered by $\mathscr{S}$. If such N does not exist, then the VC dimension is ∞.

We refer to Example 12.1.2 for an explanatory case.

Following the so-called *structural minimization principle* (Schölkopf and Smola, 2002, chapter 5), the generalization risk can be bounded by means of the empirical risk and the VC dimension of the considered class, as proved in Theorem 12.1.1, which can be found in Schölkopf and Smola (2002, Section 1.3).

Theorem 12.1.1 *Let Z be a labeled dataset that consisting of N data points, which are supposed to be sampled from an unknown probability distribution* $p(\boldsymbol{x}, y)$ *on* $\Omega \times Y$. *Consider a class* $\mathscr{S}$ *of classifying functions on* Ω, *independent of p, whose VC dimension is* $M < N$. *Then, if* $s \in \mathscr{S}$ *and* $\delta > 0$, *the bound*

$$e(f) \leq \hat{e}(Z, s) + C(M, N, \delta)$$

holds with probability $1 - \delta$, *where the so-called capacity term is*

$$C(M, N, \delta) := \sqrt{\frac{1}{N}\left(M\left(\log\frac{2N}{M} + 1\right) + \log\frac{4}{\delta}\right)}.$$

Despite more complex and sophisticated bounds being available, also by means of different capacity measures (see, e.g., Schölkopf and Smola, 2002, chapters 5 and 12, and Alon et al., 1997, and Kearns and Schapire, 1994), the presented error bound is suitable to state some important considerations concerning the learning process.

Example 12.1.2 *Hyperplanes in* $\mathbb{R}^2$ *shatter 3 noncollinear points, but no set of 4 points can be shattered (see Figure 12.1). Thus, the VC dimension of the class of separating hyperplanes in* $\mathbb{R}^2$ *is 3.*

In Theorem 12.1.1, the generalization risk in (12.2) (and thus the test error) is bounded by the sum between the empirical risk in (12.1) (i.e., the training error) and the capacity term of the class, which is monotonically increasing with the VC dimension. If we choose a *poor* class, we get a low VC dimension but a possibly high empirical risk; this situation is usually called *underfitting*. On the other hand, choosing a *rich* class, we can obtain a very small empirical risk, but the VC dimension, and thus the capacity term, is likely to be large; this condition is called *overfitting*.

Therefore, it is evident that we should look for a trade-off between minimizing the empirical risk and controlling the capacity. This is a key problem for model

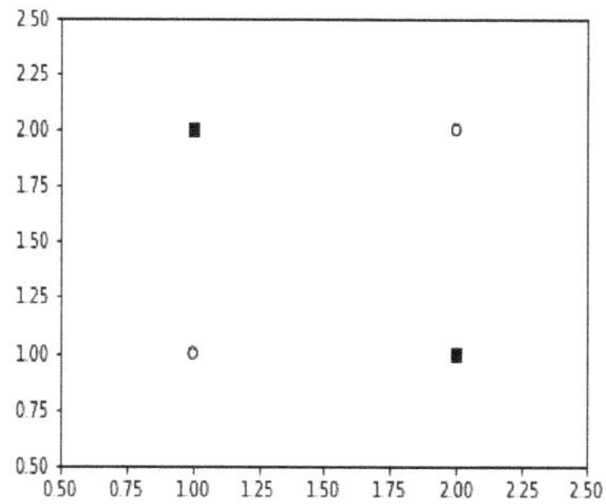

Figure 12.1 There exists no hyperplane in $\mathbb{R}^2$ that can separate the empty circles from the filled squares represented in the figure. In fact, this holds true for every set of four points in $\mathbb{R}^2$ where, taking the quadrangle built upon the considered points, the opposite elements are assigned to the same class.

selection in machine learning theory, which appears not just in classification tasks, which are discussed here, but also, for example, in regression models where $Y \subset \mathbb{R}$ (see Figure 12.2).

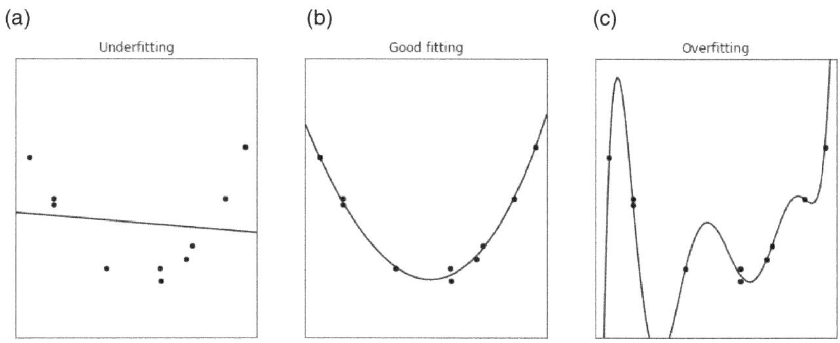

Figure 12.2 An example of model selection concerning a regression task. On (a), the chosen model is too poor and underfits the data. On (c) the data are overfitted, being the capacity of the model too large. On (b), the complexity of the model is suitable for the data at the disposal.

Many efforts have been devoted to establish how to select a suitable learning model with respect to the task considered. For example, CV (Golub et al., 1979) is a widely used empirical procedure that is adopted in order to avoid overfitting (see Section 5.2 for a detailed introduction concerning this technique).

Moreover, *regularization* strategies are dedicated to controlling the capacity of the learning class. In the presented framework, a common regularization strategy is minimizing the sum between the empirical risk and a penalty term, whose role is to penalize complex models. This strategy is considered in the context of SVMs, as we discuss in Section 13.2. For a general overview concerning regularization methods, we refer, for example, to Schölkopf and Smola (2002, chapter 4).

We point out that regularization strategies are also widely employed in regression and function approximation frameworks. In this regard, see Sections 13.1 and 9.2.5.

12.1.2 Rademacher Complexity

Here, we recall another capacity measure, namely the *Rademacher complexity* (RC) (Bartlett and Mendelson, 2002; Koltchinskii and Panchenko, 2000). Unlike the VC dimension, the RC focuses on the capacity of a class to fit random data. More precisely, instead of considering a generating probability distribution $p(\boldsymbol{x}, y)$ on $\Omega \times Y$, the RC assumes that the set of data X is sampled from a distribution $p(\boldsymbol{x})$ on Ω, while the labels are generated by considering *Rademacher variables*, which take value in $Y = \{+1, -1\}$ with equal probability. We give the definition of RC (Bartlett and Mendelson, 2002).

Definition 12.1.3 Let $X = \{\boldsymbol{x}_k,\ k = 1, \ldots, N\} \subset \Omega$ be a set of data that are supposed to be sampled from an unknown probability distribution $p(\boldsymbol{x})$ on Ω. Let us consider a class $\mathscr{S}$ of classifying functions on Ω. The empirical Rademacher complexity (ERC) is defined as

$$\hat{R}(X, \mathscr{S}) := \mathbb{E}_{\boldsymbol{\sigma}} \left[\sup_{s \in \mathscr{S}} \left| \frac{2}{N} \sum_{k=1}^{N} \sigma_i s(\boldsymbol{x}_k) \right| \middle| X \right],$$

where $\boldsymbol{\sigma} = (\sigma_1, \ldots, \sigma_N)$ is a vector of Rademacher variables and the mean value $\mathbb{E}_{\boldsymbol{\sigma}}[\cdot|X]$ averages over $\boldsymbol{\sigma}$ but keeps X fixed.
The RC of $\mathscr{S}$ is then defined as

$$R(\mathscr{S}) := \mathbb{E}_X \left[\hat{R}(X, \mathscr{S}) \right] = \mathbb{E}_{X\boldsymbol{\sigma}} \left[\sup_{s \in \mathscr{S}} \left| \frac{2}{N} \sum_{k=1}^{N} \sigma_i s(\boldsymbol{x}_k) \right| \right],$$

where the mean value $\mathbb{E}_X$ considers the distribution $p(\boldsymbol{x})$ on Ω.

We report some basic properties of ERC and RC, referring to Shawe-Taylor and Cristianini (2004, Section 4.3) and Bartlett and Mendelson (2002) for further details.

Proposition 12.1.1 *Let $X = \{\boldsymbol{x}_k,\ k = 1,\dots,N\} \subset \Omega$ be a set of data, and let $\mathscr{S}_1,\dots,\mathscr{S}_n$ be classes of classifying functions defined on Ω. Then:*

(i) If $\mathscr{S}_i \subseteq \mathscr{S}_j$, then $\hat{R}(X,\mathscr{S}_i) \leq \hat{R}(X,\mathscr{S}_j)$.
(ii) $\hat{R}(X,\bigcup_{k=1}^{N}\mathscr{S}_k) \leq \sum_{k=1}^{N}(\hat{R}(X,\mathscr{S}_k)$.
(iii) For every $c \in \mathbb{R}$, $\hat{R}(X,c\mathscr{S}) = |c|\hat{R}(X,\mathscr{S})$.

In particular, we observe that property (i) in Proposition 12.1.1 certifies that the ERC, and thus the RC, is a consistent measure of the learning capacity of a class of functions.

Remark 12.1.2 *In literature, slightly different definitions can be found where ERC and RC are not multiplied by a factor of 2 (see, e.g., the definitions of RC reported in Balcan (2011) and Kääriäinen (2004)). In all cases, ERC and RC can be seen as measures of the best correlation between functions in $\mathscr{S}$ and random data.*

A bound may be obtained for the generalization risk by means of the presented Rademacher theory, as explained in Theorem 12.1.2 (see Bartlett and Mendelson, 2002).

Theorem 12.1.2 *Let Z be a labeled dataset that consists of N labeled data, which are supposed to be sampled from an unknown probability distribution $p(\boldsymbol{x},y)$ on $\Omega \times Y$. Let us consider a class $\mathscr{S}$ of classifying functions on Ω, independent of p. Moreover, let $R(\mathscr{S})$ be the RC of the class $\mathscr{S}$. Then, if $s \in \mathscr{S}$ and $\delta > 0$, the bound*

$$e(s) \leq \hat{e}(Z,s) + \frac{R(\mathscr{S})}{2} + \sqrt{\frac{\log 1/\delta}{N}}$$

holds with probability $1-\delta$.

We remark that very similar bounds can be derived by considering the ERC in the estimate or slightly different definitions of RC (see, e.g., Balcan, 2011 and Shawe-Taylor and Cristianini, 2004, section 4.2). Moreover, see, for example, Kääriäinen (2004) for a discussion concerning the relation between the Rademacher theory and the VC dimension, along with related error bounds.

12.2 Kernels in Learning Literature

In the machine learning literature, algorithms that make use of a kernel function and of the kernel trick see below formula (12.4) are called *kernel methods*. These methods are widely employed in different applications, due to their flexibility and capability of introducing nonlinearity in preexisting linear models.

In our discussion, we consider two kernel-based models, namely kernel regularization networks (KRN) for regression in Section 13.1 and SVMs for classification in Section 13.2. Other well-known algorithms are, for example, kernel principal component analysis and kernelized k-means clustering. Kernels are also involved in Gaussian processes. We refer to Hofmann et al. (2008); Kung (2014); and Shawe-Taylor and Cristianini (2004) for a general overview and for further details concerning kernel methods.

In Section 2.1, we denoted by kernel a symmetric real-valued function κ on $\Omega \times \Omega \subseteq \mathbb{R}^{d\times d}$. This function was introduced in order to construct a data-dependent interpolation space. Then, we restricted ourselves to strictly positive definite radial kernels, in order to guarantee the well-posedness of the linear system resulting in the interpolation process (see Section 2.2.1).

Here, we start from the same definition but give a different interpretation, eventually converging to the same theoretical background presented in Chapter 2.

12.2.1 Kernels as Similarity Measure

In Problem 12.1.1, with the training dataset Z at our disposal, our aim is to construct a classifier that is capable of generalizing to a new unseen sample $\boldsymbol{x} \in \Omega$, predicting its corresponding label $y \in Y$. In other words, we want $(\boldsymbol{x}, y)$ to be in some sense *similar* to $(\boldsymbol{x}_k, y_k) \in Z$, $k = 1, \ldots, N$.
Since in Y the labels can only be identical or not, we are interested in discussing such a notion of similarity restricted to Ω. Therefore, we can introduce on $\Omega \times \Omega$ a symmetric similarity measure (or kernel)

$$\kappa : \Omega \times \Omega \longrightarrow \mathbb{R}, \quad (\boldsymbol{x}, \boldsymbol{y}) \longmapsto \kappa(\boldsymbol{x}, \boldsymbol{y}).$$

First, since $\Omega \subset \mathbb{R}^d$, we can take the Euclidean scalar product and define the kernel

$$\kappa_L(\boldsymbol{x}, \boldsymbol{y}) := \boldsymbol{x} \cdot \boldsymbol{y}, \quad \boldsymbol{x}, \boldsymbol{y} \in \Omega. \tag{12.3}$$

Indeed, it is well known that the scalar product computes the angle between two vectors, provided that they are normalized; hence, it measures a geometric similarity between data points in Ω.

Despite being simple, the linear kernel is widely adopted in the machine learning literature, being fast in calculation and competitive in many tasks, especially when dealing with very high-dimensional data Huang and Lin (2016).

On the other hand, it may be too poor to represent possible complex similarities in the data. Moreover, in applications, the data may generally consist of structured elements such as graphs or words, thus they may need to be first encoded in some dot product space.

To this aim, we consider a map

$$\hbar : \Omega \longrightarrow \widetilde{\mathscr{H}},$$

where $\widetilde{\mathscr{H}}$ is some dot product space (or pre-Hilbert space) endowed with a scalar product $\langle \cdot, \cdot \rangle_{\mathscr{H}}$. We observe that the space $\widetilde{\mathscr{H}}$ can be completed to a Hilbert space $\mathscr{H}$, by including the limit of Cauchy sequences in $\widetilde{\mathscr{H}}$.

Then, we define a similarity measure (or kernel) on $\Omega \times \Omega$ as

$$\kappa(\boldsymbol{x}, \boldsymbol{y}) = \langle \hbar(\boldsymbol{x}), \hbar(\boldsymbol{y}) \rangle_{\mathscr{H}}, \quad \boldsymbol{x}, \boldsymbol{y} \in \Omega. \tag{12.4}$$

We refer to $\hbar$ and $\mathscr{H}$ with *feature map* and *feature space*, respectively (Schölkopf and Smola, 2002, section 1.1). This flexible scheme allows the definition of kernels dedicated to words (Thorsten, 1998), graphs (Kriege et al., 2020), persistence diagrams (Guillemard and Iske, 2017), and many other structured data.

Remark 12.2.1 *It is straightforward to prove that the so-constructed κ is a positive definite kernel. Indeed, for any set of points $X = \{\boldsymbol{x}_k,\ k = 1, \ldots, N\} \subset \Omega$ and $c_1, \ldots, c_N \in \mathbb{R}$, from Definition 2.2.2, we have*

$$\sum_{i,j=1}^{N} c_i c_j \kappa(\boldsymbol{x}_i, \boldsymbol{x}_j) = \left\langle \sum_{i=1}^{N} c_i \hbar(\boldsymbol{x}_i), \sum_{j=1}^{N} c_j \hbar(\boldsymbol{x}_j) \right\rangle_{\mathscr{H}} = \left\| \sum_{i=1}^{N} c_i \hbar(\boldsymbol{x}_i) \right\|_{\mathscr{H}}^{2} \geq 0.$$

We give a further crucial interpretation for the equivalence in (12.4): If we dispose of a kernel function κ, we can actually avoid the mapping of the data points and the computation of the scalar product in the Hilbert space. This aspect, which is known as *kernel trick* in the machine learning literature, is particularly appealing from a computational point of view.

One of the most important and quoted examples is the Gaussian kernel

$$\kappa_G(\boldsymbol{x}, \boldsymbol{y}) = e^{-\gamma \|\boldsymbol{x}-\boldsymbol{y}\|^2}, \quad \boldsymbol{x}, \boldsymbol{y} \in \Omega,\ \gamma > 0, \tag{12.5}$$

where we slightly modify the notation, denoting $\gamma = \varepsilon^2$, according to a standard notation in the machine learning context.

It is possible to prove that the feature space corresponding to the Gaussian kernel is infinite dimensional (see, e.g., Shawe-Taylor and Sun, 2014). In addition to confirming the potential of the kernel trick, this example suggests that it is worthwhile to move the focus from the feature map to the kernel. More precisely, in Section 12.2.2, we present the assumptions under which evaluating a kernel function is equivalent to computing a scalar product in a Hilbert space.

12.2.2 From Kernels to Feature Spaces

In Section 12.2.1, we showed how a positive definite kernel κ can be constructed by means of a scalar product in a certain Hilbert space $\mathscr{H}$. Theorem 12.2.1 provides a complete characterization of positive definite kernels in terms of feature maps and feature spaces (Shawe-Taylor and Cristianini, 2004, section 3.2).

Theorem 12.2.1 *Let $\Omega \subseteq \mathbb{R}^d$ and let $\kappa : \Omega \times \Omega \longrightarrow \mathbb{R}$ be a kernel which is continuous or has a finite domain. Then, κ is positive definite if and only if there exists a feature map $\mathscr{h} : \Omega \longrightarrow \mathscr{H}$, where $\mathscr{H}$ is a Hilbert feature space with inner product $\langle \cdot, \cdot \rangle_{\mathscr{H}}$, such that*

$$\kappa(\boldsymbol{x}, \boldsymbol{y}) = \langle \mathscr{h}(\boldsymbol{x}), \mathscr{h}(\boldsymbol{y}) \rangle_{\mathscr{H}}, \quad \boldsymbol{x}, \boldsymbol{y} \in \Omega.$$

Remark 12.2.2 *The correspondence between kernel and feature space is not one-to-one. Indeed, while the positive definite kernel corresponding to a feature map and space is unique, on the other hand, it is possible to construct different feature maps and spaces associated with a positive definite kernel (Schölkopf and Smola, 2002, section 2.2).*
We point out that such a lack of uniqueness is usually not relevant from an applied point of view. In fact, if an algorithm is written only in terms of dot products in the feature space, it is possible to not deal with different feature maps by simply evaluating the unique kernel function, according to the presented kernel trick.

In Section 12.2.2.1, we present two different constructions for the feature map.

12.2.2.1 First Construction: RKHS Map

Referring to Section 2.3 for further details, we can consider the RKHS (see Definition 2.3.1) feature map

$$\mathscr{h} : \boldsymbol{x} \longmapsto \kappa_{\boldsymbol{x}}(\cdot), \tag{12.6}$$

where $\kappa_{\boldsymbol{x}}(\cdot) = \kappa(\boldsymbol{x}, \cdot)$, $\boldsymbol{x} \in \Omega$, and the related feature space $\mathscr{N}_{\kappa}(\Omega)$.

12.2.2.2 Second Construction: Mercer Map

Here, the construction is built on the following result (see, e.g., Schölkopf and Smola (2002, section 2.2.4) or Fasshauer and McCourt (2015, section 2.2.3)).

Theorem 12.2.2 *Let $\kappa \in \mathscr{L}_2(\Omega \times \Omega)$ be a real-valued kernel such that the integral operator* $\mathrm{T}: \mathscr{L}_2(\Omega) \longrightarrow \mathscr{L}_2(\Omega)$ *defined as*

$$(\mathrm{T}f)(\boldsymbol{x}) = \int_{\Omega} \kappa(\boldsymbol{x}, \boldsymbol{y}) f(\boldsymbol{y}) \mathrm{d}\boldsymbol{y}$$

is positive, that is

$$\int_{\Omega \times \Omega} \kappa(\boldsymbol{x}, \boldsymbol{y}) f(\boldsymbol{x}) f(\boldsymbol{y}) \mathrm{d}\boldsymbol{x} \mathrm{d}\boldsymbol{y} \geq 0.$$

Let $N_{\mathscr{H}} \in \mathbb{N} \cup \{\infty\}$ and let $(\upsilon_k)_{k=1,\ldots,N_{\mathscr{H}}} \in \mathscr{L}_2(\Omega)$ be the orthonormal eigenfunctions of the operator T, *associated to the eigenvalues $(\lambda_k)_{k=1,\ldots,N_{\mathscr{H}}} > 0$, sorted in nonincreasing order. Then, the following result holds true.*

(i) The eigenvalues $(\lambda_k)_{k=1,\ldots,N_{\mathscr{H}}}$ are absolutely summable, thus T *has finite trace.*

(ii) The expansion

$$\kappa(\boldsymbol{x}, \boldsymbol{y}) = \sum_{k=1}^{N_{\mathscr{H}}} \lambda_k \upsilon_k(\boldsymbol{x}) \upsilon_k(\boldsymbol{y})$$

holds almost everywhere on $\Omega \times \Omega$. Moreover, if $N_{\mathscr{H}} = \infty$ the series converges absolutely and uniformly almost everywhere on $\Omega \times \Omega$.

The results of Theorem 12.2.2, which can be traced back to the work of Mercer (1909), allow us to define the *Mercer feature map*

$$m : \boldsymbol{x} \longmapsto \left(\sqrt{\lambda_k} \upsilon_k(\boldsymbol{x}) \right)_{k=1,\ldots,N_{\mathscr{H}}} \tag{12.7}$$

for almost all $\boldsymbol{x} \in \Omega$. In this case, the feature space corresponds to $\ell_2^{N_{\mathscr{H}}}$, that is, the Hilbert space of possible infinite square summable sequences.

It can be proved that the positiveness of the integral operator T in Theorem 12.2.2 is equivalent to the fact that κ is positive definite (see, e.g., Schölkopf and Smola (2002, proposition 2.12)). Moreover, if $N_{\mathscr{H}} = \infty$, the uniform convergence in Theorem 12.2.2 (ii) implies that the resulting infinite dimensional feature map m can be approximated with arbitrary accuracy by a finite feature map. Finally, despite determining different feature spaces, the feature maps presented $\hbar$ and m defined in (12.6) and (12.7) can be directly related (see Schölkopf and Smola (2002, section 2.2.4) for further details).

12.2.3 Further Properties of Positive Definite Kernels

Here, we present some further properties of positive definite kernels. In particular, complex kernels can be constructed even from simple ones (Shawe-Taylor and Cristianini, 2004, section 3.4).

Proposition 12.2.1 *Let κ_1, κ_2 be positive definite kernels on $\Omega \times \Omega \subseteq \mathbb{R}^{d \times d}$. Then, the following are all positive definite kernels.*

(i) $\kappa(\boldsymbol{x},\boldsymbol{y}) = \kappa_1(\boldsymbol{x},\boldsymbol{y}) + \kappa_2(\boldsymbol{x},\boldsymbol{y})$.
(ii) $\kappa(\boldsymbol{x},\boldsymbol{y}) = a\kappa_1(\boldsymbol{x},\boldsymbol{y})$, *where* $a \in \mathbb{R}$, $a > 0$.
(iii) $\kappa(\boldsymbol{x},\boldsymbol{y}) = \kappa_1(\boldsymbol{x},\boldsymbol{y})\kappa_2(\boldsymbol{x},\boldsymbol{y})$.
(iv) $\kappa(\boldsymbol{x},\boldsymbol{y}) = f(\boldsymbol{x})f(\boldsymbol{y})$, *where* $f : \Omega \longrightarrow \mathbb{R}$.
(v) $\kappa(\boldsymbol{x},\boldsymbol{y}) = \kappa_3(t(\boldsymbol{x}),t(\boldsymbol{y}))$, *where* $t : \Omega \longrightarrow \mathbb{R}^m$, $m \in \mathbb{N}$, *and* κ_3 *is a positive definite kernel on* $\mathbb{R}^m \times \mathbb{R}^m$.
(vi) $\kappa(\boldsymbol{x},\boldsymbol{y}) = \boldsymbol{x}^\intercal \mathsf{B} \boldsymbol{y}$, *where* B *is an* $n \times n$ *positive definite matrix.*
(vii) $\kappa(\boldsymbol{x},\boldsymbol{y}) = p(\kappa_1(\boldsymbol{x},\boldsymbol{y}))$, *where* p *is a univariate polynomial with positive coefficients.*
(viii) $\kappa(\boldsymbol{x},\boldsymbol{y}) = e^{\kappa_1(\boldsymbol{x},\boldsymbol{y})}$.

The property in (v) is particularly relevant for our applications, since it is directly related to the variably scaled setting introduced in Chapter 9.

12.2.4 A Capacity Measure for Kernels

In Section 12.1, we introduced two capacity measures, namely the VC dimension and the RC. Both measures are often not easy to calculate. Here, we present a capacity measure introduced in Donini and Aiolli (2017), which is easy to calculate and is dedicated to the kernel setting.

Definition 12.2.1 Let $X = \{\boldsymbol{x}_k,\ k = 1,\ldots,N\} \subset \Omega$ be a set of data and let κ be a positive definite kernel on $\Omega \times \Omega$. Let K be the kernel matrix, such that

$$\mathsf{K}_{ij} = \kappa(\boldsymbol{x}_i,\boldsymbol{x}_j).$$

The spectral ratio (SR) of K is the ratio between its trace $\mathrm{tr}(\mathsf{K}) = \sum_{i=1}^N \mathsf{K}_{ii}$ and its Frobenius norm $\|\mathsf{K}\|_F = \sqrt{\sum_{i,j=1}^N \mathsf{K}_{ij}^2}$, that is

$$SR(\mathsf{K}) := \frac{\mathrm{tr}(\mathsf{K})}{\|\mathsf{K}\|_F}.$$

The SR is an expressiveness (or capacity) measure for kernels, in the following sense.

Definition 12.2.2 Let κ_i, κ_j be two kernels on $\Omega \times \Omega$. We say that κ_j is more expressive than κ_i whenever, for any possible dataset X, we have

$$SR(\mathsf{K}^i) \leq SR(\mathsf{K}^j),$$

where K^i and K^j are the kernel matrices on X obtained through κ_i and κ_j, respectively.

We recall some properties concerning the SR (Donini and Aiolli, 2017, section 3.1).

Proposition 12.2.2 *Let I be the $N \times N$ identity matrix and let $\mathbb{1}$ be the $N \times N$ matrix whose elements are all ones. The following properties hold.*

(i) For all kernel matrices K, we have

$$1 \leq SR(\mathsf{K}) \leq \sqrt{rank(\mathsf{K})}.$$

(ii) The matrix $\mathbb{1}$ achieves the minimum SR, that is, $SR(\mathbb{1}) = 1$.
(iii) The matrix I achieves the minimum SR ratio, that is, $SR(\mathsf{I}) = \sqrt{N}$.
(iv) The SR is invariant under multiplication with a positive scalar, that is, $SR(a\mathsf{K}) = SR(\mathsf{K})$ for any $a \in \mathbb{R}$, $a > 0$.

Finally, SR can be linked to ERC, as presented in Theorem 12.2.3 (Donini and Aiolli, 2017, subsection 3.1.2).

Theorem 12.2.3 *Let $W = \{\boldsymbol{w} = (w_1, \ldots, w_N) \in \mathbb{R}^N \mid \|\boldsymbol{w}\|_1 = 1\}$, where $\|\boldsymbol{w}\|_1 = |w_1| + \cdots + |w_N|$, and let $X = \{\boldsymbol{x}_k,\ k = 1, \ldots, N\} \subset \Omega$ be a set of data. Moreover, let κ be a positive definite kernel on $\Omega \times \Omega$ and consider the function space*

$$\mathscr{S} = \left\{ \boldsymbol{x} \longmapsto \sum_{i=1}^{N} w_i \frac{\kappa(\boldsymbol{x}, \boldsymbol{x}_i)}{\|\mathsf{K}\|_F},\ \boldsymbol{w} \in W \right\},$$

where K is the kernel matrix constructed on κ and X. Then,

$$\hat{R}(X, \mathscr{S}) \leq \frac{2}{N} \sqrt{SR(\mathsf{K})}.$$

13

Support Vector Machines and other Methods

In this chapter, we focus on two kernel-based machine learning methods. The first, that is, kernel regularization networks, can be seen as a machine learning reformulation of a regularized approximation. Then, we present support vector machines in detail, focusing on their usage in supervised classification. Finally, we briefly outline other machine learning schemes used or mentioned in the remainder of this book.

13.1 Kernel Regularization Networks

For Kernel Regularization Networks (KRN), the set of output variables consists of $Y \subset \mathbb{R}$. Indeed, KRNs are used for regression/interpolation, and they are therefore meaningful for studying, for example, the behavior of longitudinal data, that is, time series (see, e.g., Campagna et al., 2019). This allows us to extrapolate the evolution of the samples and, consequently, to make short-term predictions about the process's dynamics. For a general overview on KRNs, we refer the reader to Fasshauer and McCourt (2015) and Orr (1996).

Let then $X = \{\boldsymbol{x}_k,\ k = 1,\dots,N\} \subset \Omega$, $\Omega \subset \mathbb{R}^d$, and let $y_1,\dots,y_N \in \mathbb{R}$ be the related outputs. We aim to model the input–output relationship via a predictor $s : \Omega \longrightarrow \mathbb{R}$, which is chosen to be a linear combination of *some* kernel-based basis functions. Given a set of centers $\bar{X} = \{\bar{\boldsymbol{x}}_k,\ k = 1,\dots,\bar{N}\} \subseteq \Omega$, the KRN predictor is of the form

$$s(\boldsymbol{x}) = \sum_{k=1}^{\bar{N}} c_k \kappa(\boldsymbol{x}, \bar{\boldsymbol{x}}_k), \tag{13.1}$$

for (strictly) positive definite kernels $\kappa : \Omega \times \Omega \longrightarrow \mathbb{R}$ and for some real coefficients $c_1,\dots,c_{\bar{N}}$.

The learning function (13.1) can be interpreted as the simplest neural network, consisting of only one linear combination of basis functions, that is, a single layer.

Remark 13.1.1 *The presented regression framework shows some similarities to the interpolation setting introduced in Section 2.1. Indeed, as already observed in Section 12.1.1, regularization strategies are often introduced in the interpolation framework to improve the stability of the reconstruction process (Guastavino and Benvenuto, 2020).*
More precisely, the $\mathscr{L}_2$-regularization (Fasshauer and McCourt, 2015, chapter 14) (also called Tikhonov regularization) adopted in Section 9.2.5 for strictly positive definite radial kernels is indeed the same strategy presented here, in a more general setting, for KRNs.

The vector of coefficients $\boldsymbol{c} = (c_1, \ldots, c_{\bar{N}})^\intercal \in \mathbb{R}^{\bar{N}}$ is computed via the following minimization problem (Fasshauer and McCourt, 2015, section 18.1)

$$\min_{\boldsymbol{c}\in\mathbb{R}^{\bar{N}}} \left[\sum_{i=1}^{N} \left(y_i - \sum_{j=1}^{\bar{N}} c_j \kappa(\boldsymbol{x}_i, \bar{\boldsymbol{x}}_j) \right)^2 + \nu \sum_{j=1}^{\bar{N}} c_j^2 \right],$$

where $\nu \in \mathbb{R}_+$ is a regularization parameter.

We may take the set of kernel centers $\bar{X} \equiv X$. In that case, the kernel matrix K^Ψ of entries

$$\mathsf{K}^\Psi_{ij} = \kappa^\Psi(\boldsymbol{x}_i, \boldsymbol{x}_j), \quad i, j = 1, \ldots, N$$

is squared. Furthermore, if a strictly positive-definite kernel such as the Gaussian kernel κ_G is used, then the matrix is nonsingular. Therefore, we may look to the special setting for which $\nu = 0$. In that case, the solution can be found as

$$\boldsymbol{c} = (\mathsf{K}^\Psi)^{-1} \boldsymbol{y},$$

where $\boldsymbol{y} = (y_1, \ldots, y_N)^\intercal$.

In general, computing the inverse of the kernel matrix K might lead to serious instability issues due to its typical ill-conditioning. In Section 16.3.3, variably scaled kernels are considered to reduce such instability.

13.2 SVM for Classification

In what follows, we deal with the supervised learning Problem 12.1.1 by introducing SVMs. We restrict the discussion to the binary classification case, being the most widely adopted multiclass schemes built on multiple binary classifiers

(Bredensteiner and Bennett, 1999). For further details regarding our analysis, we refer to Schölkopf and Smola (2002, chapter 9).

We note that the SVM algorithm can be adapted for regression tasks and is a well-established method in this field. We refer to Schölkopf and Smola (2002, Section 1.6); Smola and Schölkopf (2004); or Vapnik (1995) for a complete discussion on this topic.

13.2.1 The Linear Case

We face Problem 12.1.1 geometrically. Namely, given a training dataset Z, which is built on the sets $X = \{\boldsymbol{x}_k,\ k = 1,\dots,N\} \subset \Omega$, $\Omega \subset \mathbb{R}^d$ and $Y = \{+1,-1\}$, we want to build a classifier on separating hyperplanes in $\mathbb{R}^n$. More precisely, we look for a solution to the learning problem in the class of functions

$$\mathscr{S} = \{s : \Omega \longrightarrow Y \mid s(\boldsymbol{x}) = \operatorname{sgn}(\boldsymbol{w}\cdot\boldsymbol{x}+b),\ \boldsymbol{w}\in\mathbb{R}^d,\ b\in\mathbb{R}\}.$$

13.2.1.1 The Hard Margin Setting

In the first part of this discussion, we restrict ourselves to linearly separable datasets, for which there exists at least one $s \in \mathscr{S}$ such that the training error (or empirical risk) is zero, that is, $\hat{e}(Z,s) = 0$ (see the Definition in 12.1). We depict an example of a linearly separable dataset in Figure 13.1a.

We call this setting *hard margin*, where linear separability is assumed. In this case, there are actually an infinite number of functions in $\mathscr{S}$ that achieve a training error equal to zero. Intuitively, the most attractive function is the one that is associated with the hyperplane that maximizes the margin of separation with respect to the closest element of each class, that is, the hyperplane "in the middle" between the two classes. Indeed, recalling that we are interested in classifying new samples, it seems reasonable that it has greater generalization power than the other candidates.

Remark 13.2.1 *This geometric intuition is theoretically confirmed: besides being unique, the hyperplane "in the middle" is optimal, since it can be proved that the capacity of the class $\mathscr{S}$, expressed in terms of the VC dimension (see Definition 12.1.2), decreases as the margin increases (we refer to Schölkopf and Smola, 2002, chapter 5, for a detailed discussion).* In Figure 13.1b, we show an example of an optimal hyperplane.

We formalize the problem of efficiently finding the optimal hyperplane. First, if t is the signed distance between $\boldsymbol{x} \in \Omega$ and the hyperplane defined by $\boldsymbol{w}$

and b, meaning that such a distance is negative if $\boldsymbol{w} \cdot \boldsymbol{x} + b < 0$ and nonnegative otherwise, we introduce the function

$$g(\boldsymbol{x}) := \boldsymbol{w} \cdot \boldsymbol{x} + b = t\|\boldsymbol{w}\|,$$

where $\|\cdot\|$ is the standard Euclidean norm. Indeed, $\boldsymbol{x} = \boldsymbol{x}_P + t\boldsymbol{w}/\|\boldsymbol{w}\|$, where $\boldsymbol{x}_P$ is the orthogonal projection of $\boldsymbol{x}$ in the hyperplane, which implies $g(\boldsymbol{x}_P) = 0$. Since the hyperplane does not change by varying the norm of its orthogonal vector, it is convenient to fix $t\|\boldsymbol{w}\| = 1$, which is equivalent to $t = 1/\|\boldsymbol{w}\|$. Consequently, maximizing t is equivalent to minimizing $\|\boldsymbol{w}\|$.
Finally, if now t is the distance between the hyperplane and the closest element of the training dataset, we require for every $k = 1, \ldots, N$

$$\frac{y_k g(\boldsymbol{x}_k)}{\|\boldsymbol{w}\|} \geq t.$$

Therefore, we can determine the optimal hyperplane by solving the quadratic programming problem

$$\begin{aligned} &\min_{\boldsymbol{w},b} \tfrac{1}{2}\|\boldsymbol{w}\|^2 \\ &\text{subject to } y_k(\boldsymbol{w} \cdot \boldsymbol{x}_k + b) \geq 1,\ k = 1, \ldots, N. \end{aligned} \tag{13.2}$$

The minimization task in (13.2), which is usually referred to as the hard margin SVM optimization problem in *primal form*, can be efficiently solved by several algorithms. We refer to Pallaschke and Rolewicz (1997) and Schölkopf and Smola (2002, chapter 6) for an overview of this topic.

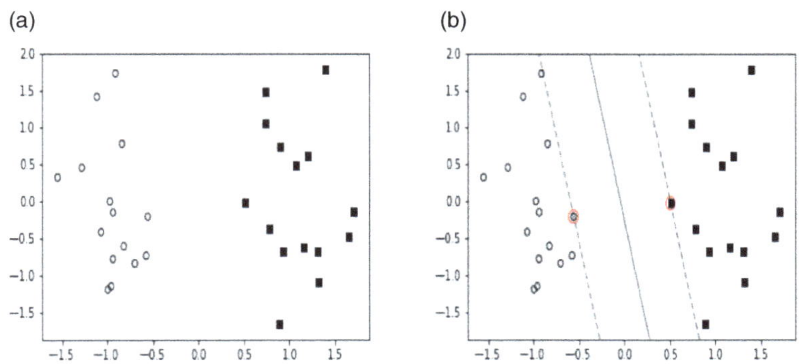

Figure 13.1 (a) An example of a linearly separable dataset in $\mathbb{R}^2$, where the elements belong to one out of two classes: empty circles or filled squares. (b) Displays the optimal separating hyperplane (solid line). The support vectors are circled in red; as highlighted by the dotted lines, they are the closest elements to the optimal hyperplane.

To take advantage of the kernel setting presented in Section 12.2, it is necessary to write the SVM optimization problem in terms of scalar products between the elements of the dataset. Observing that this is a convex optimization problem, we achieve this result by passing from the minimization of the primal form to the maximization of the *dual form* by means of the Lagrange multipliers (Fletcher, 2000, chapter 9). Namely, we consider the *Lagrangian*

$$L(\boldsymbol{w}, b, \boldsymbol{\alpha}) = \frac{1}{2}\|\boldsymbol{w}\|^2 - \sum_{k=1}^{N} \alpha_k \big(y_k(\boldsymbol{w} \cdot \boldsymbol{x}_k + b) - 1\big),$$

where $\boldsymbol{\alpha} = (\alpha_1, \dots, \alpha_N) \in \mathbb{R}^N$, $\alpha_k \geq 0$, $k = 1, \dots, N$. Then, from the *Karush–Kuhn–Tucker* (KKT) conditions (see Fletcher, 2000, Theorem 9.1.1) we obtain

$$\boldsymbol{w} = \sum_{k=1}^{N} y_k \alpha_k \boldsymbol{x}_k, \tag{13.3}$$

and the SVM optimization problem in dual form

$$\begin{aligned} &\max_{\boldsymbol{\alpha}} \textstyle\sum_{k=1}^{N} \alpha_k - \frac{1}{2} \sum_{i,j=1}^{N} y_i y_j \alpha_i \alpha_j \boldsymbol{x}_i \cdot \boldsymbol{x}_j \\ &\text{subject to } \alpha_k \geq 0,\ \textstyle\sum_{k=1}^{N} y_k \alpha_k = 0,\ k = 1, \dots, N. \end{aligned} \tag{13.4}$$

Let $\boldsymbol{\alpha}^\star = (\alpha_1^\star, \dots, \alpha_N^\star)$ be the vector of optimal parameters obtained by solving the dual problem in (13.4). Then, we get $\boldsymbol{w}^\star$ in (13.3).
Moreover, taking $\boldsymbol{x}_{sv} \in X$ so that $\boldsymbol{w} \cdot \boldsymbol{x}_{sv} + b = 1$, that is, $\boldsymbol{x}_{sv}$ is an element that realizes the minimum distance from the optimal hyperplane, we get $b^\star = 1 - \boldsymbol{w}^\star \cdot \boldsymbol{x}_{sv}$.
Therefore, we can express the optimal classifier by means of scalar products, that is

$$s(\boldsymbol{x}) = \operatorname{sgn}\left(\sum_{k=1}^{N} y_k \alpha_k^\star \boldsymbol{x} \cdot \boldsymbol{x}_k + b^\star \right).$$

Remark 13.2.2 *Another important consequence of the KKT conditions is that*

$$\alpha_k^\star \big(y_k(\boldsymbol{w}^\star \cdot \boldsymbol{x}_k + b^\star) - 1\big) = 0,\ k = 1, \dots, N.$$

In other words, $\alpha_k^\star = 0$ unless $\boldsymbol{x}_k$ realizes the minimum distance from the hyperplane. Thus, a few elements of $\boldsymbol{\alpha}^\star$ are nonzero, providing a concrete computational advantage. Such elements are called support vectors. In the hard margin setting presented, they correspond in fact to the elements of the dataset that have minimal distance with respect to the optimal hyperplane, as depicted in Figure 13.1b.

The characterization of the support vectors differs in the so-called *soft margin* scheme, which is the next step in our analysis.

13.2.1.2 From Hard to Soft Margin

Until now, we assumed linear separability, which allows the construction of separating hyperplanes that perfectly divide the two classes. However, in general, we deal with datasets that are not linearly separable. This can mainly be due to complex similarity structures or outliers, but there may also be a certain amount of noise affecting the data. We display an example in Figure 13.2a. Therefore, we usually cannot aim to find a (best) separating hyperplane, since it may not exist.

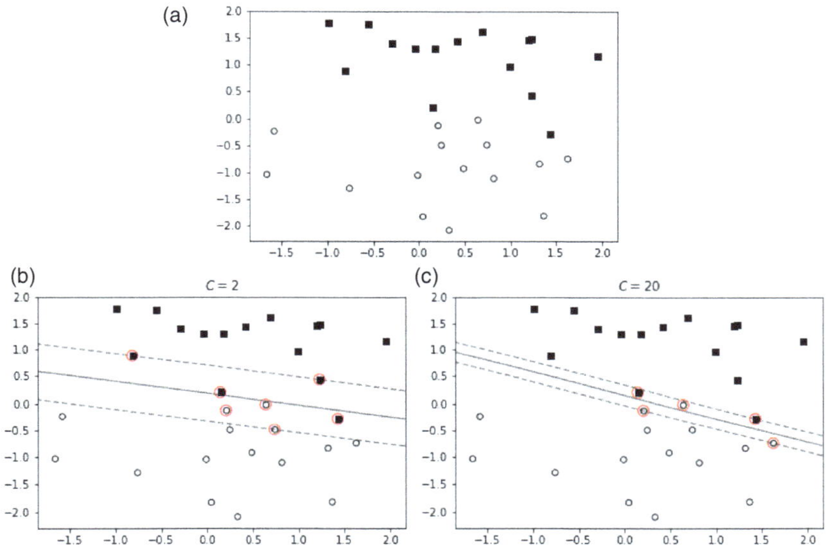

Figure 13.2 (a) An example of a nonlinearly separable dataset in $\mathbb{R}^2$, where the elements belong to one out of two classes: empty circles or filled squares. (b, c) display the results of the soft-margin SVM classification obtained with $C = 2$ (b) and $C = 20$ (c). The effects of such a hyperparameter are visible: by letting C get larger, elements on the margin are more penalized, and support vectors (circled in red) are lower in numbers.

A possible solution is to let some elements of the dataset violate the linear constraints in (13.2). To do this, we first introduce the so-called *slack variables* (Cortes and Vapnik, 1995)

$$\xi_k \geq 0, \ \xi_k \in \mathbb{R}, \ k = 1, \ldots, N.$$

Then, letting $C > 0$, $C \in \mathbb{R}$, we introduce the primal form of the *soft margin* SVM optimization problem, which is

$$\begin{aligned} &\min_{\boldsymbol{w},b} \tfrac{1}{2}\|\boldsymbol{w}\|^2 + C\textstyle\sum_{k=1}^{N} \xi_k \\ &\text{subject to } y_k(\boldsymbol{w}\cdot\boldsymbol{x}_k + b) \geq 1 - \xi_k, \ \xi_k \geq 0, \ k = 1, \ldots, N. \end{aligned} \tag{13.5}$$

Let us analyze this minimization problem by comparing it with the hard margin scheme in (13.2). The linear constraint can be relaxed using slack variables, whose sum is included in the objective function. In this way, in addition to minimizing $\boldsymbol{w}$, which controls the classifier's capacity, we also ensure that the slack variables are not substantial to avoid overfitting.

The trade-off between these two aspects is determined by the parameter C. Since such a parameter is fixed *a priori*, it is usually referred to as *hyperparameter* in the machine learning literature, and, in applications, it is generally tuned by means of CV strategies (see the discussion in Sections 12.1.1 and 5.2). In Figure 13.2b and 13.2c, we present an example of how the classifier changes according to the value of C.

Finally, by means of the same strategy adopted in the hard margin case, we get the *soft margin dual problem* (Schölkopf and Smola, 2002, section 1.5)

$$\begin{aligned}&\max_{\boldsymbol{\alpha}} \textstyle\sum_{k=1}^{N} \alpha_k - \frac{1}{2}\sum_{i,j=1}^{N} y_i y_j \alpha_i \alpha_j \boldsymbol{x}_i \cdot \boldsymbol{x}_j \\ &\text{subject to } 0 \leq \alpha_k \leq C,\ \textstyle\sum_{k=1}^{N} y_k \alpha_k = 0,\ k = 1,\ldots,N.\end{aligned} \tag{13.6}$$

Here, the only modification with respect to the hard margin scheme is that the dual variables are bounded by C; in this view, it is theoretically possible to recover the hard margin by letting $C = \infty$.

As in the primal form, it is even more evident that C controls the influence of the single elements. Moreover, it can be shown that

$$\begin{aligned}\alpha_k = 0 &\Longrightarrow y_k(\boldsymbol{w} \cdot \boldsymbol{x}_k + b) \geq 1,\\ 0 \leq \alpha_k < C &\Longrightarrow y_k(\boldsymbol{w} \cdot \boldsymbol{x}_k + b) = 1,\\ \alpha_k = C &\Longrightarrow y_k(\boldsymbol{w} \cdot \boldsymbol{x}_k + b) \leq 1.\end{aligned}$$

Therefore, in the soft-margin case, the support vectors are elements on the margin, and misclassified elements are on the wrong side of the hyperplane; see Figure 13.2b and 13.2c for an example.

13.2.2 The Nonlinear Case

In many situations, hyperplanes can be insufficient for building a performing classifier. In some cases, this may happen even if the underlying data structure is not elaborated, but a linear separator is not geometrically convenient. A valuable strategy is to look for a function $\hbar$ such that the data mapped through $\hbar$ becomes well treatable by a linear separator, as in Example 13.2.1.

Focusing on the flexible soft margin scheme, the formalization of such a procedure only requires some adjustments with respect to the optimization algorithm in (13.5). Indeed, the nonlinear version of the SVM algorithm in primal form is

$$\begin{array}{l} \min_{\boldsymbol{w},b} \frac{1}{2}\|\boldsymbol{w}\|^2 + C\sum_{k=1}^{N} \xi_k \\ \text{subject to } y_k(\boldsymbol{w}\cdot \hbar(\boldsymbol{x}_k)+b) \geq 1-\xi_k,\ \xi_k \geq 0,\ k=1,\ldots,N. \end{array} \tag{13.7}$$

Moreover, the decision function becomes

$$s(\boldsymbol{x}) = \operatorname{sgn}\left(\boldsymbol{w}^\star \cdot \hbar(\boldsymbol{x}) + b^\star\right),$$

being $\boldsymbol{w}^\star$, $b^\star$ the optimal parameters obtained.

Example 13.2.1 *Let us consider the dataset in Figure 13.3a. Although it is clear that a circumference could perfectly represent and separate the data, such a decision boundary is not achievable using the presented linear constructions. Then, for* $(x_1,x_2) \in \mathbb{R}^2$ *we take the map* $\hbar : \mathbb{R}^2 \longrightarrow \mathbb{R}^3$ *defined as*

$$\hbar(x_1,x_2) := \left(x_1, x_2, \sqrt{x_1^2+x_2^2}\right). \tag{13.8}$$

The resulting mapped dataset, which is displayed in Figure 13.3c, is indeed linearly separable in $\mathbb{R}^3$*. There, we then construct a classifier using the discussed SVM algorithm, that is, by solving the problem defined in* (13.6)*; this is shown in Figure 13.3d. Finally, we can project back from* $\mathbb{R}^3$ *to* $\mathbb{R}^2$*, obtaining then a nonlinear classifier in the original domain, as depicted in Figure 13.3b.*
Here, the kernel function related to (13.8) *in* $\mathbb{R}^{2\times 2}$ *is*

$$\kappa(\boldsymbol{x},\boldsymbol{y}) = x_1y_1 + x_2y_2 + \sqrt{(x_1^2+x_2^2)(y_1^2+y_2^2)}.$$

If the function $\hbar$ maps the data to a very high-dimensional feature space, the primal form in (13.7) is computationally inconvenient. In contrast, passing to the dual form, we obtain the objective function

$$\max_{\boldsymbol{\alpha}} \sum_{k=1}^{N} \alpha_k - \frac{1}{2}\sum_{i,j=1}^{N} y_i y_j \alpha_i \alpha_j \hbar(\boldsymbol{x}_i)\cdot \hbar(\boldsymbol{x}_j).$$

In this case, we can take advantage of the kernel trick (see 12.4) and avoid the computation of the possible high-dimensional scalar product in the feature space. Therefore, if κ is the positive definite kernel associated to the feature map $\hbar$, then the nonlinear SVM optimization problem in dual form is

$$\begin{array}{l} \max_{\boldsymbol{\alpha}} \sum_{k=1}^{N} \alpha_k - \frac{1}{2}\sum_{i,j=1}^{N} y_i y_j \alpha_i \alpha_j \kappa(\boldsymbol{x}_i,\boldsymbol{x}_j). \\ \text{subject to } 0 \leq \alpha_k \leq C,\ \sum_{k=1}^{N} y_k\alpha_k = 0,\ k=1,\ldots,N. \end{array} \tag{13.9}$$

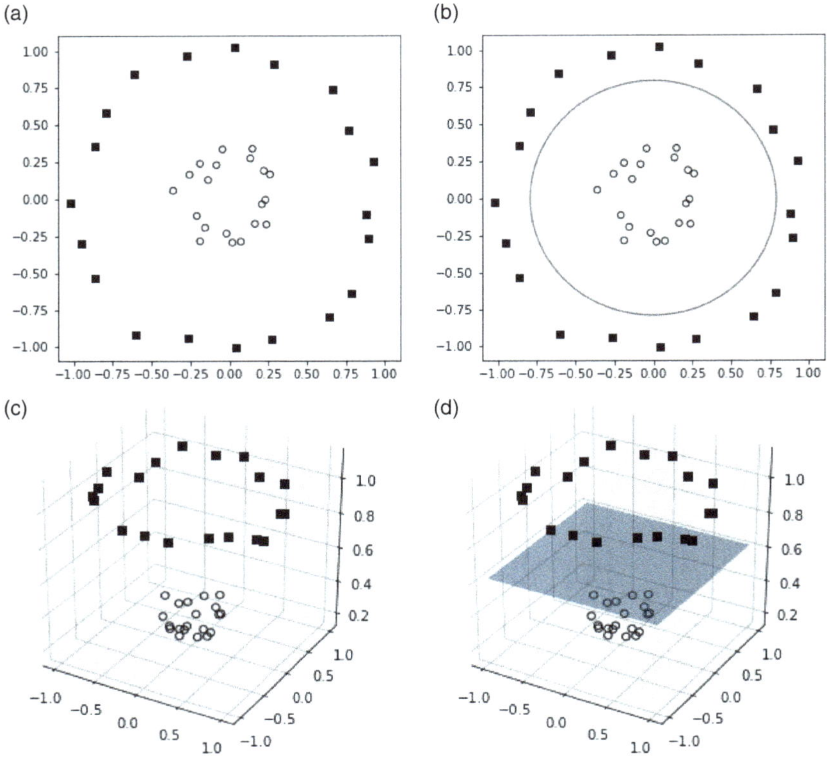

Figure 13.3 (a–d) The discussed construction of a nonlinear SVM classifier out of a suitable feature map and kernel.

Moreover, proceeding as in the linear case, we finally obtain the classifier

$$s(\boldsymbol{x}) = \operatorname{sgn}\left(\sum_{k=1}^{N} y_k \alpha_k^\star \kappa(\boldsymbol{x}_i, \boldsymbol{x}_j) + b^\star \right).$$

Among all the SVM optimization problems presented, the one in (13.9) is actually a general framework that includes all the others. Indeed, the linear case can be recovered simply by considering the linear kernel in (12.3).

As discussed at the end of Section 12.2.1, in general, we take the kernel function directly, without constructing it via a feature map. In particular, widely considered choices are the linear kernel and the Gaussian kernel defined in (12.3) and (12.5), respectively.

Regarding the Gaussian kernel, in Figure 13.4 we present an example that highlights the effects of different values of the hyperparameters C and γ on the construction of the SVM classifier. Although the role of the hyperparameter C has already been depicted in Figure 13.2, in Figure 13.4 we observe the effects of different combinations of C and γ. As the value of γ increases, according to the properties of the Gaussian function, we obtain more localized classifiers that are more dependent on the singular elements.

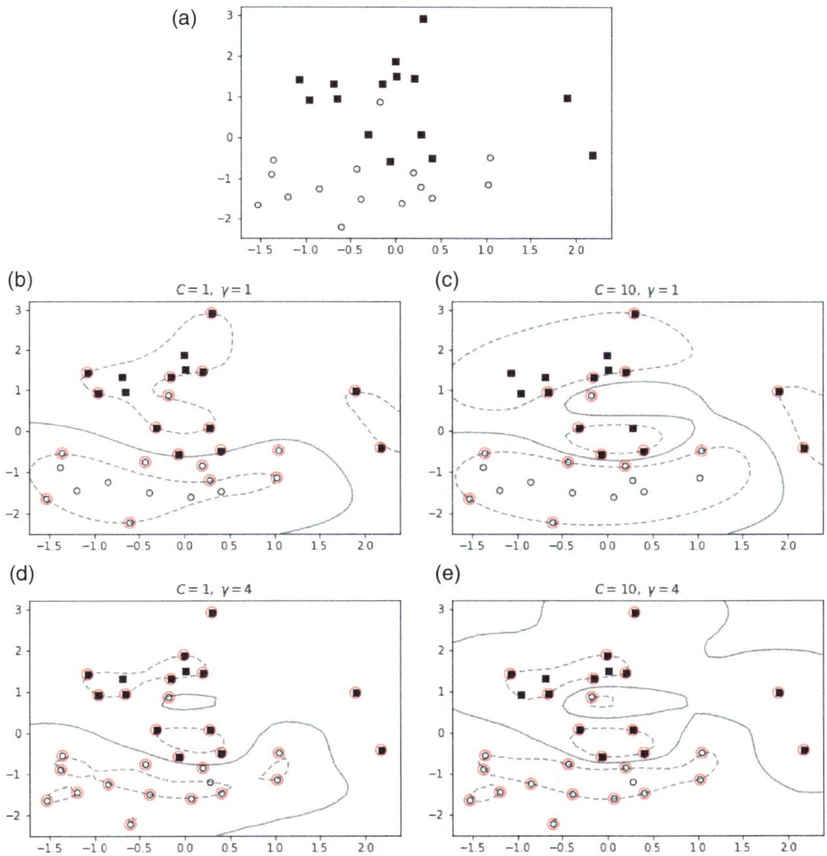

Figure 13.4 (a–e) Given a dataset, we train the SVM algorithm taking the Gaussian kernel and considering different values of the hyperparameters, which are displayed on the top of every figure. As previously done, we use a solid line for the decision boundary and circle the support vectors in red.

13.3 A Brief Note on Some Other Classifiers

Here, we provide brief words on classification methods considered or mentioned in the following that are not kernel-based by nature. In particular, we focus on the ability of such classifiers to provide *probabilistic* outputs.

As a general setting, let $\Omega \subset \mathbb{R}^d$ and let $X = \{\boldsymbol{x}_k,\ k = 1,\dots,N\} \subset \Omega$ be the set of the input data. We consider the setting learning Problem 12.1.1 in Chapter 12, but here we extend to the multiclass framework, that is $Y = \{C_1,\dots,C_\ell\}$, $\ell \in \mathbb{N}$. Especially with the VSK setting, we are interested in the *probability values*

$$P_j(\tilde{\boldsymbol{x}}) := P(\tilde{\boldsymbol{x}} \in C_j|\tilde{\boldsymbol{x}}),\ j = 1,\dots,\ell\ ,$$

where $\tilde{\boldsymbol{x}} \in \Omega$ is a test sample to be classified.

13.3.1 Naive Bayes

The Naive Bayes (NB) classifier (Aggarwal, 2014; Maron, 1961) is a simple method that is based on the well-known Bayes Theorem. It is called "naive" because its strong assumption is that all features are mutually independent. Nevertheless, it is easily implementable and widely adopted in many applications.

Let $\tilde{\boldsymbol{x}} = (\tilde{x}_1,\dots,\tilde{x}_d) \in \Omega$. The NB classifier computes

$$P_j(\tilde{\boldsymbol{x}}) = \frac{\prod_{i=1}^d P(\tilde{x}_i|C_j)P(C_j)}{P(\tilde{\boldsymbol{x}})},$$

classifying

$$\mathcal{C}(\tilde{\boldsymbol{x}}) = \text{argmax}_{j=1,\dots,\ell} P_j(\tilde{\boldsymbol{x}}).$$

The *likelihood* $\prod_{i=1}^d P(\tilde{x}_i|C_j)$, the *prior* $P(C_j)$, and the *evidence* $P(\tilde{\boldsymbol{x}})$ depend on a fixed probability distribution (e.g., Gaussian or Bernoulli), whose parameters are typically estimated from the data at the disposal. It is possible to avoid computing the evidence by normalizing.

In particular, we recall the Gaussian likelihood (Ren et al., 2009)

$$P(\tilde{x}|C_j) = \frac{1}{\sqrt{2\pi(\sigma_j)^2}} e^{-\left(\frac{\tilde{x}-\mu_j}{\sqrt{2}\sigma_j}\right)^2},$$

where μ_j and σ_j are the mean and standard deviation related to C_j, respectively.

13.3.2 *k*-Nearest Neighbors

Given a metric (e.g., Euclidean), the k-nearest neighbors (kNN) classifier (Altman, 1992) is a machine learning method that classifies a new test instance by simply looking at the class of the closest training elements. More precisely, given the value of the hyperparameter $k \in \mathbb{N}$, the test sample is assigned to the most common class among the k closest training elements. It is preferable to set k to an odd value to avoid ties.

Such a method is easy to comprehend and interpret. As a drawback, it may be computationally expensive if the number of training elements is large. A probabilistic output can be obtained from a kNN classifier by taking into consideration the respective proportions of the classes related to the nearest elements. This is described in Figure 13.5, where we present an example that also highlights the possible decisive role of the hyperparameter k, which is usually tuned by means of CV.

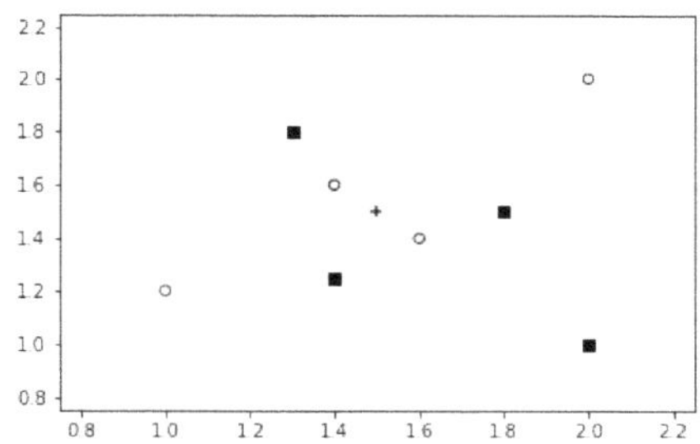

Figure 13.5 Let C_1 be the class of empty circles, C_2 the class of black squares. Moreover, let $\tilde{\boldsymbol{x}}$ be a test sample to be classified (the black plus). If $k = 3$, $P_1(\tilde{\boldsymbol{x}}) = 2/3$, $P_2(\tilde{\boldsymbol{x}}) = 1/3$, and $\tilde{\boldsymbol{x}}$ is assigned to C_1. On the other hand, if $k = 5$ we have $P_1(\tilde{\boldsymbol{x}}) = 2/5$, $P_2(\tilde{\boldsymbol{x}}) = 3/5$, and $\tilde{\boldsymbol{x}}$ is assigned to C_2.

13.3.3 Random Forest

A random forest (RF) (Breimann, 2001; Ho, 1995) is an ensemble method that consists of a collection of decision trees. In every tree, decision rules are inferred from the training set. More precisely, such laws are structured as a flowchart that starts at the *root* node and ends at various final nodes, also called *leaves*.

In principle, the decision tree algorithm selects the splitting rules in such a way that the overall *entropy* related to the output classes is minimized in the consequent layer. Intuitively, such entropy is low at a given node when the output class is ubiquitous among the included training elements relative to other courses. We refer to Rokach and Maimon (2014) for a complete overview of different versions of the decision tree algorithm. In Figure 13.6, we display an example of a decision tree.

In an RF, every tree votes for the classification of a test instance; therefore, an immediate probabilistic output is achievable as discussed for the kNN classifier.

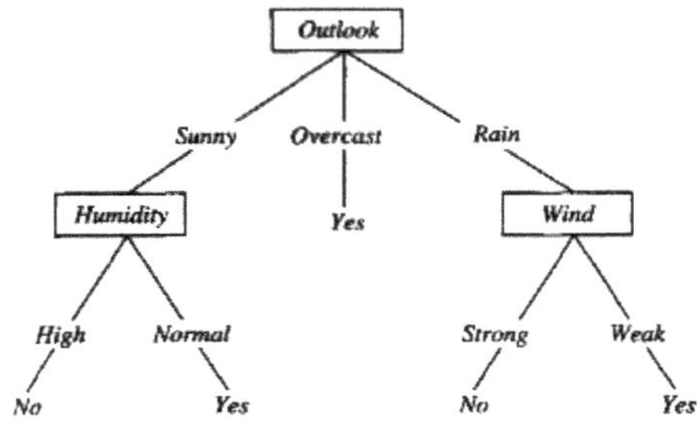

Figure 13.6 An example of decision tree from Mitchell (1997, figure 3.1). Here, the task is to decide whether or not to play tennis according to the weather.

13.3.4 Neural Network

Artificial neural networks (NNs) are a flexible and powerful machine learning method, initially inspired by biological NNs. In the last decades, a lot of effort has been put into the research of various NN configurations that can be adapted to a wide range of tasks (Wenzel et al., 2026). Although their learning capacity enables them to tackle many complex tasks, they often require substantial amounts of training data and lack interpretability. For a complete overview, we refer the reader to Goodfellow et al. (2016).

Here, we present some details about the structure of a deep fully connected (FC) feed-forward NN.

A set of artificial neurons is organized in an *input* layer, an *output* layer, and *hidden* layers, which in the FC setting are also denoted as *dense* layers. The m neurons $z_1^j, \ldots, z_m^j$ of the layer j transmit their values to the next layer $j+1$ through weighted synapses and a chosen activation function a. More precisely, the value of the neuron z_i^{j+1} is given by

$$z_i^{j+1} = a\left(\sum_{k=1}^{m} w_{k,j}^i z_k^j + w_0^j\right)$$

where $w_{k,j}^i$ is the weight related to the synapse that connects z_k^j to z_i^{j+1} and w_0^j is the bias related to the layer j.

In hidden layers, it is common to adopt the ReLU (rectified linear unit) activation function, which is defined as

$$\mathrm{ReLU}(x) = \max(0, x).$$

For the output layer, we use the softmax activation function, which is a suitable choice for multiclass classification problems. Indeed, it is a generalization of the logistic function, and it is defined as

$$\mathrm{softmax}(\boldsymbol{x}) = \frac{1}{\bar{e}}(e^{x_1}, e^{x_2}, \dots, e^{x_n}),$$

where $\boldsymbol{x} = (x_1, x_2, \dots, x_d)$ is a d-dimensional vector and $\bar{e} = \sum_{j=1}^{n} e^{x_j}$. We point out that such a choice guarantees an immediate probabilistic output, that is, the prediction $\mathcal{C}(\boldsymbol{x}, \boldsymbol{w})$ is an ℓ-dimensional vector that can be interpreted as a probability distribution over the ℓ classes. We display the structure of such an NN in Figure 13.7.

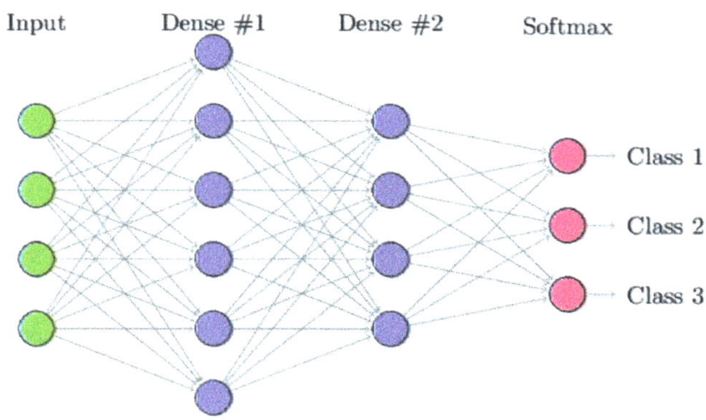

Figure 13.7 An example of a deep feed-forward fully connected neural network from Pelletier et al. (2019). Here, the number of output classes is $\ell = 3$.

In the training process, for all elements $\boldsymbol{x}_i \in X$, $i = 1, \dots, N$, the weights of the network are adjusted to minimize a loss function $L(\mathcal{C}(\boldsymbol{x}_i, \boldsymbol{w}), y_i))$, where $\mathcal{C}(\boldsymbol{x}_i, \boldsymbol{w})$ is the prediction of the network given the input $\boldsymbol{x}_i$ and weights $\boldsymbol{w}$ (including bias terms), and y_i is the actual output related to $\boldsymbol{x}_i$, which in this context is *one-hot* encoded (see, e.g., Harris and Harris, 2013), therefore it is a ℓ-dimensional vector whose coordinates are all zero except the one representing the actual output class, which is set to the value of 1.

In this context, a common choice is to adopt the *cross-entropy* loss

$$L(\mathcal{C}(\boldsymbol{x}_i, \boldsymbol{w}), y_i) = \sum_{k=1}^{\ell} y_i^k \log(\mathcal{C}^k(\boldsymbol{x}_i, \boldsymbol{w})),$$

where y_i^k and $\mathcal{C}^k(\boldsymbol{x}_i, \boldsymbol{w})$ denote the k-th coordinate of the vectors y_i and $\mathcal{C}(\boldsymbol{x}_i, \boldsymbol{w})$, respectively. In the optimization process, the *Adam* first-order gradient-based optimizer introduced by Kingma and Ba (2015) represents a common choice; we refer to Goodfellow et al. (2016) for a detailed treatment.

14
Local-to-global SVMs

In Section 13.2, we introduced SVMs, which are one of the most popular classification methods (Schölkopf and Smola, 2002). However, similarly to kernel-based approximation, both the computational complexity and the memory requirements may become prohibitive as the number of instances grows; see, for example, Menon (2009) for a general overview. Consequently, many strategies have been proposed in literature to overcome this issue. Several works address approximating the kernel matrix and reducing the number of support vectors to produce rather sparse global classifiers (Maji et al., 2013; Joachims and Yu, 2009). Another class of approaches aims to construct a global classifier by gluing and exploiting *many* local algorithms, considering, for example, local kNN classifiers, weights, or indicator functions (Blanzieri and Melgani, 2006; Collobert et al., 2001; Dekel and Shamir, 2012; Ding et al., 2017; Jose et al., 2013; Wang et al., 2010; Zhang et al., 2014). A desirable property of a properly defined *local-to-global classifier* is to be able to classify a new example, also taking into account the contributions of neighboring classifiers, not relying on a single local method. With this idea in mind, in this chapter, we investigate what we call the local-to-global SVM (LGSVM) method, a global classifier obtained by solving several *small* SVM classification problems, which is based on the PU approach presented in Chapter 7. Nevertheless, as we shall see, some modifications are required when moving to the classification framework. For a complete treatment including more details and tests, we refer to Marchetti and Perracchione (2022b).

14.1 A PU Approach for Classification

The first adaptation that we consider restricts the set of possible coverings of our domain to those that possess some necessary requirements.

Definition 14.1.1 Let $\Omega \subset \mathbb{R}^d$ and let $\mu : \Omega \times \Omega \longrightarrow \mathbb{R}$ be a semi-metric (or semi-distance) and let Ω be connected and bounded in the μ semi-metric. Moreover, let $Z_N = \{(\boldsymbol{x}_i, y_i) : \boldsymbol{x}_i \in X_N,\ y_i = \mathcal{C}_1, \ldots, \mathcal{C}_\ell,\ i = 1, \ldots, N,\ \ell \in \mathbb{N},\ \ell \geq 2\} \subset \Omega$, be a dataset whose elements belong to one of the ℓ classes. Let us consider a collection of sets (or patches) $\mathcal{D}_p := \{\Omega_1, \ldots, \Omega_p\}$, where $\Omega_k \subset \mathbb{R}^d$ is open, connected, and bounded with respect to μ for each $k = 1, \ldots, p$. Then $\mathcal{D}_p$ is an admissible covering for (Ω, X_N) if the following properties are satisfied.

1. The set $\mathcal{D}_p$ is a covering of Ω, i.e., $\Omega \subseteq \bigcup_{k=1}^{p} \Omega_k$ (covering assumption).
2. For each $k = 1, \ldots, p$, there exists at least one point $\boldsymbol{x} \in \Omega$ such that $\boldsymbol{x} \in \Omega_k$ (partitioning assumption).
3. In each patch Ω_k, $k = 1, \ldots, p$, at least two classes of $\mathcal{D}_1, \ldots, \mathcal{C}_\ell$ are represented by some elements of X_N (representation assumption).

Moreover, $\mathcal{C}_p$ is a quasi-admissible covering for (Ω, X_N) if the representation assumption does not hold true.

Note that the representation assumption is introduced to accommodate the classification setting. On the other hand, the following proposition indicates that every set of a *(quasi-)admissible covering* has to present some overlap with its neighboring patches, and it is a direct consequence of the partitioning assumption.

Proposition 14.1.1 *Let $\mathcal{D}_p = \{\Omega_1, \ldots, \Omega_p\}$ be a (quasi-)admissible covering for (Ω, X_N). Then, if $p \geq 2$, for every patch Ω_i there exists a set Ω_j, $i, j = 1, \ldots, p$, $i \neq j$, such that $\Omega_i \cap \Omega_j \neq \varnothing$.*

Proof. It is sufficient to prove the case $p = 2$. By virtue of the partitioning assumption in Definition 14.1.1, there exist $\boldsymbol{x}, \boldsymbol{y} \in \Omega$ such that $\boldsymbol{x} \in \Omega_1$ and $\boldsymbol{y} \in \Omega_2$. Then, letting $g : [0,1] \longrightarrow \Omega$ be a continuous path from $\boldsymbol{x} = g(0)$ to $\boldsymbol{y} = g(1)$ in Ω, which exists since Ω is connected, there exists $\lambda^\star$ such that $\boldsymbol{\zeta} = g(\lambda^\star) \notin \Omega_1$ is on the boundary of Ω_1. If $\boldsymbol{\zeta} \notin \Omega_2$, then $\boldsymbol{\zeta} \notin \Omega$. Therefore, $\boldsymbol{\zeta} \in \Omega_2$ holds true. Moreover, since Ω_2 is open, there exists a neighborhood of $\boldsymbol{\zeta}$ contained in Ω_2. On the other hand, since $\boldsymbol{\zeta}$ is on the boundary of Ω_1, such a neighborhood necessarily contains elements that belong to Ω_1. Hence, $\Omega_1 \cap \Omega_2 \neq \varnothing$. □

On a (quasi-)admissible covering, we can consider a family of compactly supported functions which form a PU. In particular, we can consider the Shepard weights defined in (7.1). Here, the role of such compactly supported functions is weighting and assembling the contributions of the local classifiers.

14.2 The LGSVM Classifier

We first deal with the construction of this method when an admissible covering for (Ω, X_N) is at the disposal, which is the case where a local classifier can be defined on each subdomain. Moreover, without losing generality, we restrict ourselves to the binary classification setting with $\ell = 2$, where $\mathcal{C}_1 = \{y = -1\}$ and $\mathcal{C}_2 = \{y = +1\}$. Indeed, the LGSVM method generalizes to the multiclass framework according to the standard SVM algorithm.

Let $\Omega \subset \mathbb{R}^d$, Z_N be a training dataset and let $\mathcal{D}_p$ be an admissible covering for (Ω, X_N) according to Definition 14.1.1. With some abuse of notation, from now on, we denote the subdomains by $\Omega_k := \Omega \cap \Omega_k$, $k = 1, \dots, p$. First, in order to construct the LGSVM classifier, different SVMs are trained separately and locally on each subdomain. More precisely, letting N_k be the number of training elements in Ω_k, $\boldsymbol{x}_i^{(k)} \in X_{N_k} = X_N \cap \Omega_k$ and $y_i^{(k)}$ the associated label, with $i = 1, \dots, N_k$, we solve (13.9) on each subdomain Ω_k. We then obtain a local classifier of the form

$$s_k(\boldsymbol{x}) = \sum_{i=1}^{N_k} \alpha_k^i y_i^{(k)} \kappa(\boldsymbol{x}, \boldsymbol{x}_i^{(k)}) + b_k, \quad \boldsymbol{x} \in \Omega, \tag{14.1}$$

for each $k = 1, \dots, p$. Then, the *local-to-global* step is performed by means of the PU structure, that is, the *global* decision function of the LGSVM classifier is given by

$$\mathcal{s}(\boldsymbol{x}) = \text{sign}\left(\sum_{k=1}^{p} s_k(\boldsymbol{x}) \omega_k(\boldsymbol{x})\right),$$

where ω_k is the function defined in (7.1) and s_k is the hyperplane calculated in (14.1). As proved in Marchetti and Perracchione (2022b, proposition 2), the LGSVM classifier can effectively solve many subproblems by providing local information to the global classifier.

The representation assumption in Definition 14.1.1 may be too restrictive in concrete situations, where the presence of one-class clusters in the data distribution is common. Therefore, we present a possible extension of the LGSVM setting to quasi-admissible coverings, which implies that we need to face the problem of managing the case of one-class subdomains, for which local SVM classifiers are not definable. Moreover, we need to deal with empty patches, that is, patches that do not contain any training element. Let us suppose that $\widetilde{p}$ patches satisfy the representation assumption, where $\widetilde{p} \in \mathbb{N}$, $\widetilde{p} \leq p$, and write $\widetilde{\Omega}_j$ in place of Ω_j if it satisfies such assumption. Then we define the union of such subdomains

$$\mathscr{M} := \bigcup_j \widetilde{\Omega}_j, \tag{14.2}$$

and we refer to it as *admissible subcovering*, which might consist of disjoint unions of overlapping sets. On the admissible subcovering, we construct the LGSVM classifier as explained earlier. On the other hand, if a test example $\bar{\boldsymbol{x}}$ to be classified is such that $\bar{\boldsymbol{x}} \in \Omega \setminus \mathscr{M}$, we face two possibilities:

1. The test element $\bar{\boldsymbol{x}}$ is contained in one-class patches. In this case, we assign $\bar{\boldsymbol{x}}$ to the class that characterizes the subdomains it belongs to. This is a natural choice that preserves the *local purposes* of the presented method.
2. The test element $\bar{\boldsymbol{x}}$ lies on empty subdomains. Here, we classify it by means of the *fast nearest centroid* (NC) classifier Tibshirani et al. (2002). Note that other auxiliary classifiers might be taken into account.

14.3 Constructing the Covering

While in the classical PU setting the metric used for decomposing the domain usually coincides with the Euclidean distance, here we allow the more general case of patches built via a given semi-metric $\mu : \Omega \times \Omega \longrightarrow \mathbb{R}$. Let $r \in \mathbb{R}^+$. We consider coverings where each patch Ω_k is an open μ-ball, that is, it is characterized by a center $\hat{\boldsymbol{x}}_k \in \Omega$ and a radius r, that is, $\Omega_k = \{\boldsymbol{x} \in \Omega : \mu(\hat{\boldsymbol{x}}_k, \boldsymbol{x}) \leq r\}$. Particularly, given $\boldsymbol{x}, \boldsymbol{y} \in \Omega$, we consider the standard Euclidean metric $\mu_{\mathrm{E}}(\boldsymbol{x}, \boldsymbol{y}) = \|\boldsymbol{x} - \boldsymbol{y}\|_2$, and on the *cosine semi-metric* $\mu_{\mathrm{CS}}(\boldsymbol{x}, \boldsymbol{y}) = 1 - \mathrm{CS}(\boldsymbol{x}, \boldsymbol{y})$, where the so-called Cosine Similarity (CS) is defined as

$$\mathrm{CS}(\boldsymbol{x}, \boldsymbol{y}) = \frac{\boldsymbol{x} \cdot \boldsymbol{y}}{\|\boldsymbol{x}\|_2 \|\boldsymbol{y}\|_2}.$$

In Figure 14.1, given a set of training instances, we give an illustrative example of subdomains constructed as μ-balls centered on μ-grids covering the domain $[-1, 1]^2$. Precisely, we compare the Euclidean metric and cosine semi-metric. The presented *grid* approach where the covering is built upon μ-grids might be not effective for datasets having *many* features, as it may result to be computationally demanding or even infeasible. As a solution, we can consider a *sparse* approach, where the purpose of covering the entire domain is given up. Instead, letting $p \in \mathbb{N}$, $p \geq 1$ be the number of centers and $r > 0$ be a fixed μ-radius, we choose the centers among the training samples at the disposal. With respect to the grid approach, which is directly linked to classical PU from the approximation context, with this sparse strategy it is not trivial to guarantee some *a*

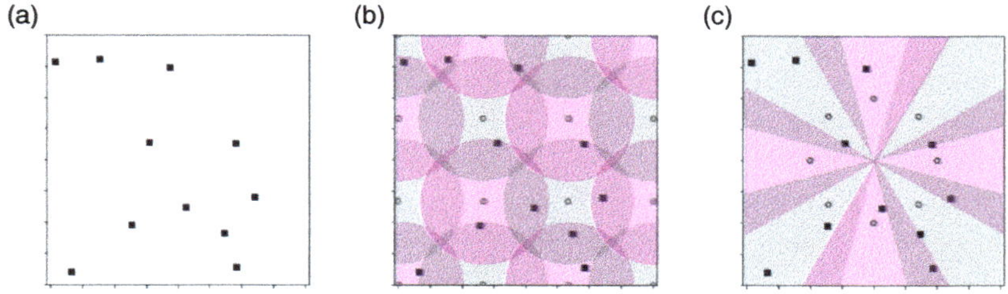

Figure 14.1 (a) The dataset (black squares). (b) We consider a grid of 16 centers (empty circles) and Euclidean balls with radius $r = 0.5$, while in (c) we take eight centers on a circumference and we decompose the domain by means of balls of radius $r = 1 - \cos(\pi/6)$ in the cosine semi-metric.

priori overlap among patches, and a preliminary tuning of the parameters p and r in a validation step is recommended. On the other hand, since test instances are usually *close to* some training data in the high-dimensional space, we can expect to classify a large amount of test samples without taking patches in empty regions of the domain. In Figure 14.2, we provide an illustrative example of subdomains construed as μ-balls via the sparse approach on $[-1, 1]^2$. The presented LGSVM method is summarized in Algorithm 14.3.1.

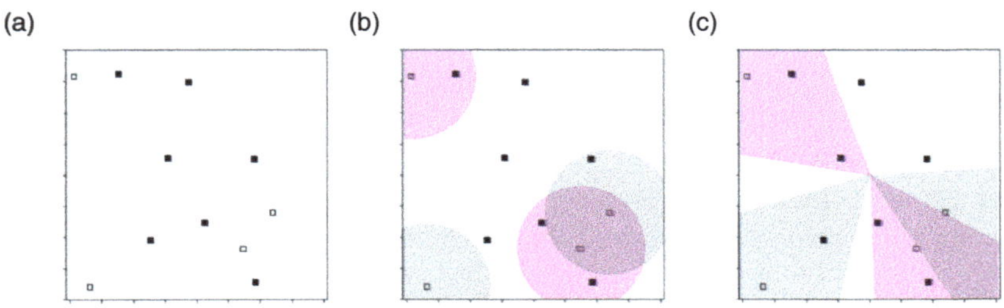

Figure 14.2 The sparse approach. (a) We choose four centers (empty squares) among the dataset of Figure 14.1a. (b) We use Euclidean balls with $r = 0.5$, while in (c) we consider balls of radius $r = 1 - \cos(\pi/6)$ in the cosine semi-metric.

Algorithm 14.3.1 (LGSVMs)

`INPUTS`: ⋆ *The training dataset* $Z_N = \{(\boldsymbol{x}_i, y_i),\ i = 1, \dots, N\} \subset \Omega \times \{-1, +1\}$;

⋆ *a set of* μ*-balls* $\mathscr{D}_p = \{\Omega_1, \dots, \Omega_p\}$;

⋆ *the set* $\mathscr{M}$ *as in* (14.2);

⋆ *an unknown test sample* $\tilde{\boldsymbol{x}} \in \Omega$;

`Grid approach`: $\mathscr{D}_p$ *is a quasi-admissible covering for* (Ω, X_N);

`Sparse approach`: $\Omega_1, \dots, \Omega_p$ *are centered at p distinct training elements respectively.*

`OUTPUTS`: *The predicted label* $\tilde{y} \in \{-1, +1\}$ *for* $\tilde{\boldsymbol{x}}$.

`Case` $\tilde{\boldsymbol{x}} \in \mathscr{M}$: *The element* $\tilde{\boldsymbol{x}}$ *is classified via* $\tilde{y} = \mathscr{s}(\tilde{\boldsymbol{x}}) = sign\left(\sum_{k=1}^{\bar{p}} s_k(\tilde{\boldsymbol{x}})\omega_k(\tilde{\boldsymbol{x}})\right)$.

`Case` $\tilde{\boldsymbol{x}} \notin \mathscr{M}$: `if` $\tilde{\boldsymbol{x}}$ *is contained in one-class balls only, then* $\tilde{y} = y_i$.

`else` *the label* $\tilde{y}$ *is assigned via the auxiliary classifier.*

14.4 LGSVMs Implementation and Results

The computational advantage provided by LGSVMs with respect to classical SVMs is determined by the efficiency provided by the PU scheme, which has already been highlighted in Chapter 7. In this section, we give a flavor of the efficiency, in seconds, and effectiveness, in terms of computed f_1-score (i.e., the harmonic mean between precision and recall), of the presented LGSVMs; for a detailed exposition, we refer to Marchetti and Perracchione (2022b).

In Figure 14.3, we report some results obtained in classification tests involving 2D simulated datasets and the 3D *skin segmentation dataset* Bhatt and Dhall (n.d.). In Figure 14.4, we restrict to the sparse approach as we consider the higher dimensional *adult dataset* Kohavi (1996) and *swarm behavior dataset* available at the link `https://archive.ics.uci.edu/ml/datasets/Swarm+Behaviour`. LGSVMs can retain the effectiveness of SVMs but are more efficient. However, especially with the sparse approach, a proper tuning of the involved parameter is important to get *optimized* results.

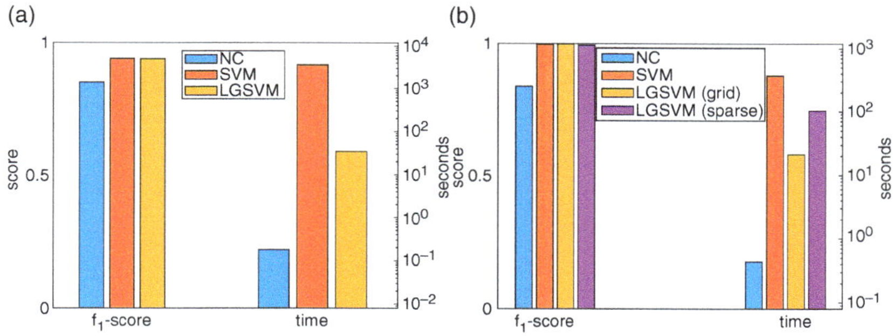

Figure 14.3 (a) Simulated 2D tests ($k = 5{,}00{,}000$). We display the cases of linear SVM and LGSVM (grid approach) with the cosine semi-metric. (b) Skin segmentation dataset. We display the cases of Gaussian SVM and LGSVM computed with the grid approach ($p = 146$), and the sparse setting ($p = 80$ and $r = 0.1$). In both cases, we take the cosine semi-metric.

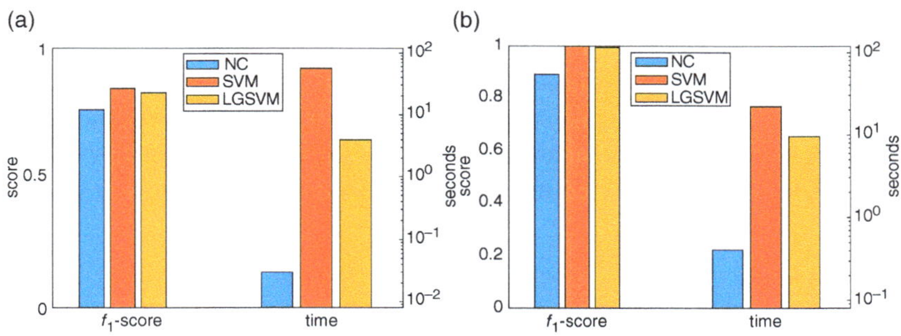

Figure 14.4 (a) Adult dataset. We display the cases of Gaussian SVM and linear LGSVM with parameters $p = 100$, $r = 0.4$. (b) Swarm dataset. We display the cases of linear SVM and LGSVM with parameters $p = 2000$, $r = 0.4$.

15
Learning Data-Driven Optimized Kernels

Standard kernels with scalar shape parameters lack the flexibility to capture anisotropic data structures in high dimensions. We present two-layered (2L) kernels based on a matrix transformation that learns optimal feature directions. For Matérn kernels, we present convergence rates that depend on the effective dimension of this matrix, with faster rates when the null space aligns with invariant subspaces of the target function. The 2L-KOGA algorithm combines kernel optimization via stochastic gradient descent with greedy node selection through VKOGA. Numerical experiments on regression benchmarks show that 2L-KOGA outperforms standard kernels when significant dimensions can be reduced, with performance correlating to the concentration of singular values.

15.1 Shape Matrices

As previously discussed, an RBF kernel usually depends on a single length scale or shape parameter ε. However, especially in dimensions $d \gg 1$, different directions within the data, not necessarily aligned with Euclidean ones, might be unequally relevant. In order to gain more flexibility, in Chapter 9, we presented VS(D)Ks, where, however, a scaling function needs to be accurately designed. In this chapter, we consider a more general parameters-based approach by taking a $b \times d$ matrix $\mathsf{A}_{\boldsymbol{\theta}}$

$$\mathsf{A}_{\boldsymbol{\theta}} = \begin{pmatrix} \theta_1^1 & \dots & \theta_1^d \\ \vdots & \ddots & \vdots \\ \theta_b^1 & \dots & \theta_b^d \end{pmatrix},$$

and then considering the kernel

$$\kappa_{\boldsymbol{\theta}}(\boldsymbol{x}, \boldsymbol{y}) = \kappa(\mathsf{A}_{\boldsymbol{\theta}}\boldsymbol{x}, \mathsf{A}_{\boldsymbol{\theta}}\boldsymbol{y}).$$

For the special choice of $\mathsf{A}_{\boldsymbol{\theta}} = \varepsilon \cdot \mathsf{I}_d$, where I_d is the $d \times d$ identity matrix, we obtain the classical RBF setting, while if $\mathsf{A}_{\boldsymbol{\theta}} = \mathrm{diag}(\varepsilon_1, \dots, \varepsilon_d)$ we recover the so-called anisotropic kernels. In the following, we focus on the $d \times d$ case, that is $b = d$, as we will motivate later on (see Remark 15.3.1).

First, we point out that the resulting model can be intended and visualized as a common NN: two hidden layers, where the first layer is used to map the input evaluation point $\boldsymbol{x}$ through the matrix $\mathsf{A}_{\boldsymbol{\theta}}$; see Figure 15.1.

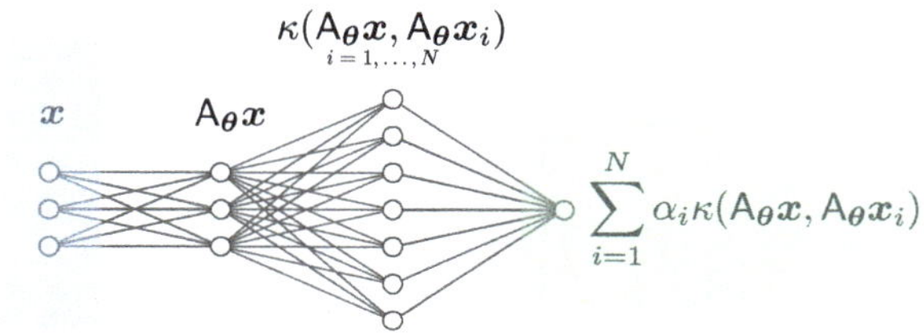

Figure 15.1 Example of a 2L kernel machine.
Visualization of the evaluation at $\boldsymbol{x}$ of the presented kernel machine. The input layer is highlighted in blue (left), while the output layer is framed in green (right). In this example, $b = d = 3$ and $N = 7$.

In fact, this is not just a matter of visualization, because the purpose is to optimize this 2L kernel machine by leveraging tools from deep learning. Before we dive into this discussion, we will first analyze some theoretical properties of $\kappa_{\boldsymbol{\theta}}(\boldsymbol{x}, \boldsymbol{y})$. For a detailed treatment of all the aspects treated in this chapter, we refer to Wenzel et al. (2024a).

15.2 Theoretical Analysis of *Optimized* Kernels

The kernel $\kappa_{\boldsymbol{\theta}}$ can be naturally understood as a 2L kernel according to the deep kernel representer theorem (Bohn et al., 2019) . Consequently, if $\{\boldsymbol{x}_1, \dots, \boldsymbol{x}_N\} \subset \mathbb{R}^d$ is such that the data matrix $[\boldsymbol{x}_1, \dots, \boldsymbol{x}_N] \in \mathbb{R}^{d \times N}$ has rank d, then the optimal approximation in the corresponding RKHS can be obtained through a linear combination of at most N basis functions for each layer. Therefore, thanks to this finite representation, $\kappa_{\boldsymbol{\theta}}$ can be treated as a classical kernel κ for our purposes.

In order to understand the impact of the first kernel layer on the performance of the overall kernel machine, we analyze the matrix $\mathsf{A}_{\boldsymbol{\theta}}$ via its singular value decomposition given by

$$\mathsf{A}_{\boldsymbol{\theta}} = \mathsf{U}\Sigma\mathsf{V}^\intercal$$

with orthogonal matrices $\mathsf{U},\mathsf{V} \in \mathbb{R}^{d\times d}$, respectively, consisting of the left and right singular vectors and the diagonal matrix $\Sigma \in \mathbb{R}^{d\times d}$ with the nonnegative singular values on the diagonal. Denoting the columns of U and V by, respectively, $\boldsymbol{u}_i$ and $\boldsymbol{v}_i, i = 1,\dots,d$ we have

$$\mathsf{A}_{\boldsymbol{\theta}}\mathsf{V} = \mathsf{U}\Sigma \quad \Leftrightarrow \quad \mathsf{A}_{\boldsymbol{\theta}}\boldsymbol{v}_i = \sigma_i \boldsymbol{u}_i \quad \forall i = 1,\dots,d.$$

For a given input $\boldsymbol{x} \in \mathbb{R}^d$, which can be decomposed as $\boldsymbol{x} = \sum_{i=1}^d \langle \boldsymbol{x}, \boldsymbol{v}_i \rangle_{\mathbb{R}^d} \boldsymbol{v}_i$ we obtain

$$\begin{aligned}
\mathsf{A}_{\boldsymbol{\theta}}\boldsymbol{x} &= \sum_{i=1}^d \langle \boldsymbol{x}, \boldsymbol{v}_i \rangle_{\mathbb{R}^d} \mathsf{A}_{\boldsymbol{\theta}}\boldsymbol{v}_i = \sum_{i=1}^d \langle \boldsymbol{x}, \boldsymbol{v}_i \rangle_{\mathbb{R}^d} \sigma_i \boldsymbol{u}_i \\
\Rightarrow \mathsf{A}_{\boldsymbol{\theta}}(\boldsymbol{x} - \tilde{\boldsymbol{x}}) &= \sum_{i=1}^d (\langle \boldsymbol{x}, \boldsymbol{v}_i \rangle_{\mathbb{R}^d} - \langle \tilde{\boldsymbol{x}}, \boldsymbol{v}_i \rangle_{\mathbb{R}^d}) \sigma_i \boldsymbol{u}_i \\
\Rightarrow \|\mathsf{A}_{\boldsymbol{\theta}}(\boldsymbol{x} - \tilde{\boldsymbol{x}})\|_{\mathbb{R}^d}^2 &= \sum_{i=1}^d (\langle \boldsymbol{x}, \boldsymbol{v}_i \rangle_{\mathbb{R}^d} - \langle \tilde{\boldsymbol{x}}, \boldsymbol{v}_i \rangle_{\mathbb{R}^d})^2 \sigma_i^2.
\end{aligned} \tag{15.1}$$

From (15.1) we see that the distance is scaled along the directions provided by the right singular vectors of $\mathsf{A}_{\boldsymbol{\theta}}$ by the corresponding singular value $\sigma_i \geq 0$. In particular, if a singular value σ_i is very small or even zero, this means that data along the directions of the corresponding right singular vectors $\boldsymbol{v}_i$ are squeezed or even mapped to the same point. This is a crucial observation for the convergence analysis, which can be derived by considering $\kappa_{\boldsymbol{\theta}}$ as a standard RBF kernel that acts on the transformed domain $\mathsf{A}_{\boldsymbol{\theta}}\Omega \subset \mathbb{R}^{\mathrm{rank}(\mathsf{A})}$. We distinguish two cases: $\mathrm{rank}(\mathsf{A}_{\boldsymbol{\theta}}) = d$ and $\mathrm{rank}(\mathsf{A}_{\boldsymbol{\theta}}) < d$. For the first case, Theorem 15.2.1 shows that we obtain at least the same rate of convergence as if one used the standard kernel κ instead of the 2L kernel $\kappa_{\boldsymbol{\theta}}(\boldsymbol{x},\boldsymbol{y})$. For the second case, Theorem 15.2.2 shows that the convergence rate is indeed increased, depending on the number of singular values of the matrix $\mathsf{A}_{\boldsymbol{\theta}}$ equal to zero. To avoid technical discussions, we will present these results restricting to the class of Matérn kernels (see 4.6). If the domain Ω is bounded and Lipschitz, their native spaces are equivalent to Sobolev spaces.

Theorem 15.2.1 *Consider the Matérn RBF kernel $\kappa = \phi_M^\beta$ (see* 4.6*), $\beta > d/2$, on a bounded Lipschitz domain $\Omega \subset \mathbb{R}^d$. Consider $f \in \mathscr{N}_\kappa(\Omega)$ and the kernel interpolant s_{X_N} using the 2L kernel $\kappa_{\boldsymbol{\theta}}$ with* $\mathrm{rank}(\mathsf{A}_{\boldsymbol{\theta}}) = d$.
Then the following pointwise error estimate holds:

$$|f(\boldsymbol{x}) - s_{X_N}(\boldsymbol{x})| \leq C h_{X_N,\Omega}^{\beta - d/2}.$$

For asymptotically equidistributed points, i.e., $h_{X_N,\Omega} \asymp N^{-1/d}$, it then holds

$$|f(\boldsymbol{x}) - s_{X_N}(\boldsymbol{x})| \leq CN^{1/2-\beta/d}.$$

Proof. We consider the 2L kernel $\kappa_{\boldsymbol{\theta}}$ as a standard Matérn RBF kernel κ applied to the transformed data $\mathsf{A}_{\boldsymbol{\theta}}X_N$ from the transformed domain $\mathsf{A}_{\boldsymbol{\theta}}\Omega$. As we have $\mathrm{rank}(\mathsf{A}_{\boldsymbol{\theta}}) = d$, we have $\dim(\mathsf{A}_{\boldsymbol{\theta}}\Omega) = \dim(\Omega) = d$. By standard Sobolev arguments (as $x \mapsto \mathsf{A}_{\boldsymbol{\theta}}x$ is just a linear full rank transformation) we have $\mathscr{W}_2^{\beta}(\Omega) \asymp \mathscr{W}_2^{\beta}(\mathsf{A}_{\boldsymbol{\theta}}\Omega)$, in particular $f \circ A_{\boldsymbol{\theta}}^{-1} \in \mathscr{W}_2^{\beta}(\mathsf{A}_{\boldsymbol{\theta}}\Omega)$. Furthermore, due to $\mathrm{rank}(\mathsf{A}_{\boldsymbol{\theta}}) = d$ all the singular values of $\mathsf{A}_{\boldsymbol{\theta}}$ are positive, that is, it holds $\sigma_{\min}(\mathsf{A}_{\boldsymbol{\theta}})\|x\|_2 \leq \|\mathsf{A}_{\boldsymbol{\theta}}x\|_2 \leq \sigma_{\max}(\mathsf{A}_{\boldsymbol{\theta}})\|x\|_2$ or in short $\|\cdot\|_2 \asymp \|\mathsf{A}_{\boldsymbol{\theta}} \cdot\|_2$. Therewith we obtain for the respective fill distances

$$\begin{aligned}
h_{\mathsf{A}_{\boldsymbol{\theta}}X_N,\mathsf{A}_{\boldsymbol{\theta}}\Omega} &\equiv \sup_{\tilde{\boldsymbol{x}} \in \mathsf{A}_{\boldsymbol{\theta}}\Omega} \left(\min_{\tilde{\boldsymbol{x}}_k \in \mathsf{A}_{\boldsymbol{\theta}}X_N} \|\tilde{\boldsymbol{x}} - \tilde{\boldsymbol{x}}_k\|_2 \right) \\
&= \sup_{\boldsymbol{x} \in \Omega} \left(\min_{\boldsymbol{x}_k \in X_N} \|\mathsf{A}_{\boldsymbol{\theta}}\boldsymbol{x} - \mathsf{A}_{\boldsymbol{\theta}}\boldsymbol{x}_k\|_2 \right) \\
&\asymp \sup_{\boldsymbol{x} \in \Omega} \left(\min_{\boldsymbol{x}_k \in X_N} \|\boldsymbol{x} - \boldsymbol{x}_k\|_2 \right) = h_{X_N,\Omega}.
\end{aligned}$$

Hence, we can make use of the error bound (4.8) to derive the final statement.
□

For a detailed proof of Theorem 15.2.2, we refer the interested reader to Wenzel et al. (2024a, Theorem 3.4).

Theorem 15.2.2 *Consider the Matérn RBF kernel $\kappa = \phi_M^{\beta}$ (see 4.6), $\beta > d/2$, on a bounded Lipschitz domain $\Omega \subset \mathbb{R}^d$. Consider $f \in \mathscr{N}_{\kappa}(\Omega)$ and the kernel interpolant s_{X_N} using the 2L kernel $\kappa_{\boldsymbol{\theta}}$ such that* $\mathrm{rank}(\mathsf{A}_{\boldsymbol{\theta}}) =: d_{\mathit{eff}} < d$.
Assume that $f \in \mathscr{N}_{\kappa}(\Omega)$ is invariant along the subspace $\mathrm{Null}(\mathsf{A}_{\boldsymbol{\theta}}) \subset \mathbb{R}^d$, *that is, $f(\boldsymbol{x}) = f(\boldsymbol{x}')$ for any $\boldsymbol{x}, \boldsymbol{x}' \in \Omega$ with $\boldsymbol{x} - \boldsymbol{x}' \in$* $\mathrm{Null}(A_{\boldsymbol{\theta}})$. *For points $X_N \subset \Omega$ such that $h_{\mathsf{A}_{\boldsymbol{\theta}}X_N,\mathsf{A}_{\boldsymbol{\theta}}\Omega} \asymp N^{-1/d_{\mathit{eff}}}$ it then holds*

$$|f(\boldsymbol{x}) - s_{X_N}(\boldsymbol{x})| \leq CN^{\frac{d}{2d_{\mathit{eff}}} - \frac{\beta}{d_{\mathit{eff}}}}.$$

Remark 15.2.1 *Note that $\frac{d}{2d_{\mathit{eff}}} - \frac{\beta}{d_{\mathit{eff}}} < \frac{1}{2} - \frac{\beta}{d} < 0$ due to $\beta > d/2$, that is, the decay rate is faster than in Theorem 15.2.1.*

Theorem 15.2.2 can be used in the following way. Given $f \in \mathscr{N}_{\kappa}(\Omega)$, which is invariant in some directions, if we can choose $\kappa_{\boldsymbol{\theta}}$ in such a way that the null space $\mathrm{Null}(\mathsf{A}_{\boldsymbol{\theta}})$ of the matrix $\mathsf{A}_{\boldsymbol{\theta}}$ coincides with this invariant subspace of f, then we can obtain faster convergence with respect to a standard kernel. In the next section, we discuss how we can effectively choose $\kappa_{\boldsymbol{\theta}}$ in this sense.

15.3 Parameters Optimization

The idea we explore now is to *learn* the kernel $\kappa_{\boldsymbol{\theta}}$, which consists in learning the entries of the matrix $\mathsf{A}_{\boldsymbol{\theta}}$, therefore interpreting the 2L kernel machine as a NN. As briefly outlined in Section 13.3.4, it is necessary to define a proper loss function L, so that the training (or optimization) procedure consists of

$$\min_{\mathsf{A}_{\boldsymbol{\theta}} \in \Theta} L(X_N, F_N, \mathsf{A}_{\boldsymbol{\theta}})$$

being Θ the parameters' space. In this framework, the loss consists of the CV error (see Section 5.2.2). However, instead of directly evaluating the CV error at X_N with respect to F_N using the kernel $\kappa_{\boldsymbol{\theta}}$, we evaluate the k-fold CV error on so-called *mini-batches*, which are randomly drawn subsets $(X_{\text{batch}}, F_{\text{batch}})$ of input data with corresponding target values of size $n_{\text{batch}} \ll N$ from the large data set X_N with corresponding target data F_N. This is common in the deep learning context, where it can be referred to as stochastic learning. In addition to classical deep learning tools (Goodfellow et al., 2016; Kingma and Ba, 2015), here we can also take advantage of the ERA for a faster computation of the CV error (see Section 5.2). More technical details on kernel machine's optimization and implementation can be found in the referring paper Wenzel et al. (2024a).

Remark 15.3.1 *Autoencoders (Goodfellow et al., 2016) are another well-established machine learning method in which data are mapped onto a feature space whose dimensionality is typically smaller than the original. However, autoencoder models are often designed as deep networks composed by many parameters, and they are deeply trained and tailored on the data at disposal, while the discussed kernel machine is built to be an auxiliary light linear model, that may enhance the performance of the subsequent approximation process. Moreover, by setting $b = d$, the possible reduction in dimensionality is not set a priori, but is accommodated during the training process.*

15.4 Classification Tests with 2L-KOGA

In Wenzel et al. (2024a), the so-called 2L-KOGA algorithm combines first the optimization and usage of the optimized kernel $\kappa_{\boldsymbol{\theta}}$, then a f-greedy nodes selection is performed on top of the optimized kernel through VKOGA (see 8.1). We now present some results from Wenzel et al. (2024a) to highlight the effectiveness of 2L-KOGA in possible very high-dimensional settings. In

Table 15.1, we outline the machine learning datasets considered, which are benchmarks for regression tasks.

Table 15.1 Overview on the considered machine learning data sets

Short name	Number of examples	Number of features
ct	53500	379
sgemm	241600	14
wecs	72000	48
mlr_knn_rng	111753	132
fried	40768	10
kegg_undir_uci	64608	27
online_video	68784	26
diamonds	53940	29
stock	59049	9
sarcos	44484	21
query_agg_count	200000	4
road_network	434874	2

In Figures 15.2 and 15.3, we compare the results obtained with 2L-KOGA to the ones achieved by considering a standard VKOGA where a Matérn kernel was used with 10 logarithmically equally spaced shape parameters ε between 0.05 and 10. Note that the results are the average of five runs for each dataset, in order to guarantee some statistical robustness.

As evident from the displayed results, 2L-KOGA obtains a better performance with respect to classical kernels in the experiments in Figure 15.2, while in Figure 15.3 the situation is way different. In order to investigate on this observation, for each dataset we can take a look at the singular values of the optimized matrices $\mathsf{A}_{\boldsymbol{\theta}}$ by considering the *cumulative power function*

$$m \mapsto \frac{\sum_{i=1}^{m} |\sigma_i(\mathsf{A}_{\boldsymbol{\theta}})|}{\sum_{i=1}^{d} |\sigma_i(\mathsf{A}_{\boldsymbol{\theta}})|}, \quad 1 \leq m \leq d,$$

whereby the singular values $\sigma_i(\mathsf{A}_{\boldsymbol{\theta}})$ are ordered from the largest to the smallest according to their absolute value, allowing us to understand how much power is clustered in the top singular values. This is shown in Figure 15.4.

Together with Figures 15.2 and 15.3, one can see that the 2L-KOGA approach works better when the ratio of the singular approaches the value one more quickly, that is, relevant active directions are caught by $\mathsf{A}_{\boldsymbol{\theta}}$ and many other subspaces are not active (very small related singular values). This agrees with the theoretical analysis provided in Section 15.2.

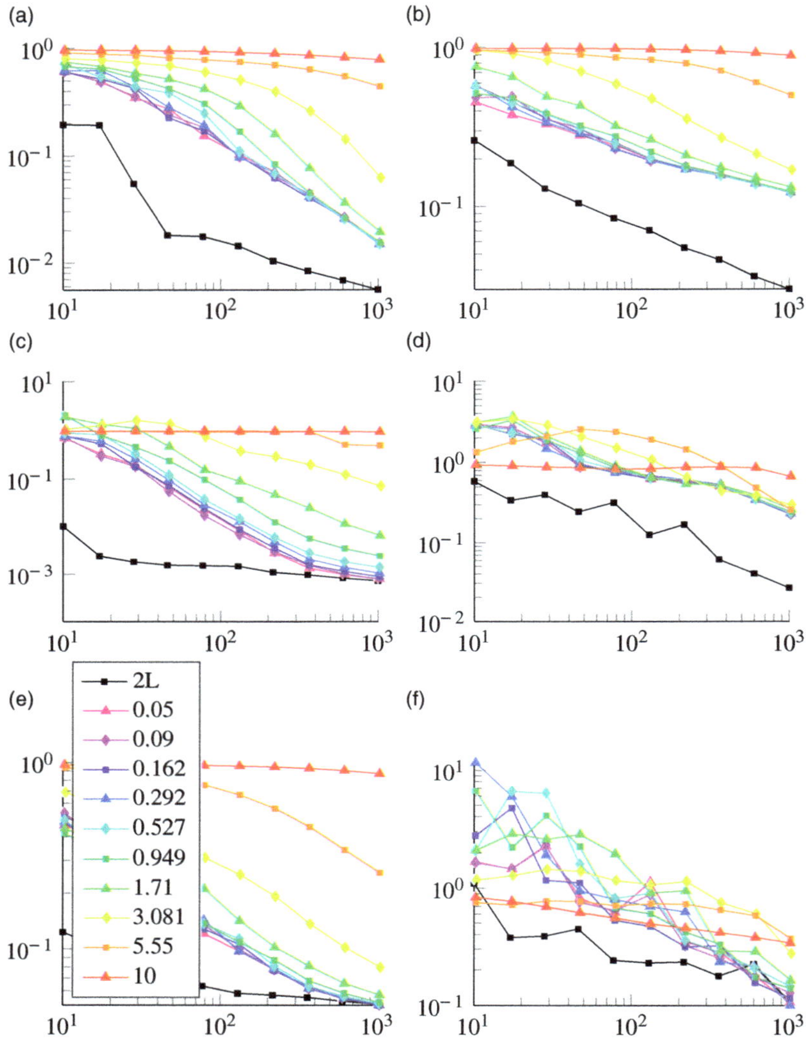

Figure 15.2 Visualization of the test MSE (y-axis) over the number of greedily selected centers (x-axis) for the first six datasets (a) ct; (b) sgemm; (c) wecs; (d) mlr_knn_rng; (e) fried; (f) kegg_undir_uci. The black line shows the 2L-KOGA, while the colored lines show the use of standard kernels with different length scale parameters ε.

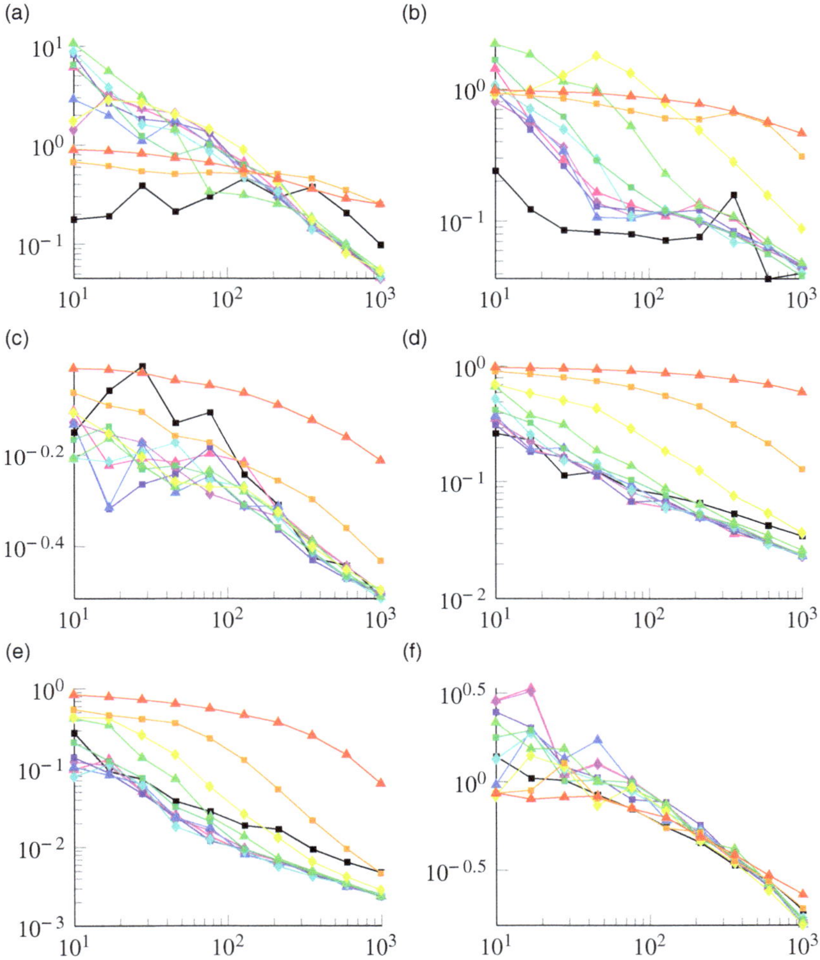

Figure 15.3 Visualization of the test MSE (y-axis) over the number of greedily selected centers (x-axis) for the second six datasets (a) online_video; (b) diamonds; (c) stock; (d) sarcos; (e) query_agg_count; (f) road_network. The black line (squared) shows the 2L-KOGA, while the colored lines show the use of standard kernels with different length scale parameters ε.

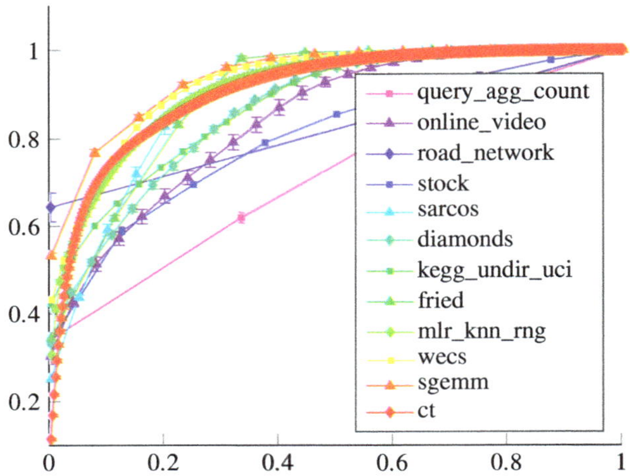

Figure 15.4 Visualization of the ratio $\sum_{i=1}^{m} |\sigma_i(\mathsf{A}_{\boldsymbol{\theta}})| / \sum_{i=1}^{d} |\sigma_i(\mathsf{A}_{\boldsymbol{\theta}})|$ (mean and standard deviation over five reruns) (y-axis) over the ratio m/d of considered singular values (x-axis).

16

Choosing and Learning the Scaling Function for the VSKs

In this chapter, we resume the thread of VSKs (see Chapter 9), and we deepen some crucial aspects of the variably scaled setting, also by taking advantage of learning techniques.

As already empirically shown, the efficacy of interpolating via VSKs depends on the definition of a *proper* scaling function. Still, no general numerical recipes for constructing it have been provided. Chapters 9, 10, and 11 suggest that the scaling function for the VSKs should mimic the target function, but no theoretical justification is provided, except for the case of discontinuous target functions. In what follows, we fill both the gaps: We prove that a scaling function reflecting the target one may lead to enhanced approximation accuracy, and we provide a user-independent tool for learning the scaling function by means of δNN, i.e., NNs capable of dealing with possible discontinuities (Della Santa and Pieraccini, 2023).

To address these issues, we demonstrate, using Lebesgue functions, that a scaling function that reflects the behavior of the target can lead to improved approximation accuracy. As a further confirmation of our claims, the scaling function returned by the δNNs confirms the theoretical findings: the learned scaling function closely resembles the target. This data-driven approach, namely the δNN–VSKs in what follows, offers a user-independent, adaptive solution for straightforwardly defining VSKs (Audone et al., 2025).

To make the presentation clearer, we want to stress the fact that the scheme used to find the nearly optimal scaling function ψ, based on the so-called δNNs, could be thought of as an independent step of the analysis carried out in Section 16.1. Indeed, the use of δNNs is primarily due to numerically supporting the claim that the scaling function should resemble the target one, even if we do not have theoretical evidence about the fact that such a choice is definitely the best one. Nevertheless, independently of the results reported in Section 16.1, the δNN algorithm returns a learned function ψ that definitely has the same key features of the target.

Finally, we discuss some further extensions of the variably scaled setting and further effective choices of the scaling function.

16.1 Investigating the Scaling Function for VSKs

In case of discontinuities, intuitively, if the scaling function ψ is chosen so that it has discontinuities at the same edges of f, so is for the radial kernel κ^{Ψ} and hence for s^{Ψ}. Following such intuition, we are interested in proposing bounds that highlight the dependence of the error on the scaling function and in providing an algorithm to automatically define such a scaling function.

As already pointed out in Chapter 6, for any set of distinct points $\{\boldsymbol{x}_i\}_{i=1}^N$, there exist cardinal functions $u_j^{\Psi} \in \text{span}\{\kappa^{\Psi}(\cdot,\boldsymbol{x}_j), j=1,\dots,N\}$ so that the VSK interpolant can be written as

$$s^{\Psi}(\boldsymbol{x}) = \sum_{j=1}^{N} f_j u_j^{\Psi}(\boldsymbol{x}),$$

being $u_j^{\Psi}(\boldsymbol{x}) := u_j((\boldsymbol{x},\psi(\boldsymbol{x}))$. Such functions are computed by solving the linear system

$$\mathsf{K}\boldsymbol{u}^{\Psi} = \boldsymbol{\kappa}^{\Psi},$$

where $\boldsymbol{u}^{\Psi} = (u_1((\boldsymbol{x},\psi(\boldsymbol{x})),\dots,u_N((\boldsymbol{x},\psi(\boldsymbol{x})))^{\mathsf{T}}$ and the vector on the right-hand side is given by $\boldsymbol{\kappa}^{\Psi} = (\kappa((\boldsymbol{x},\psi(\boldsymbol{x})),(\boldsymbol{x}_1,\psi(\boldsymbol{x}_1))),\dots,\kappa((\boldsymbol{x},\psi(\boldsymbol{x})),(\boldsymbol{x}_N,\psi(\boldsymbol{x}_N))))^{\mathsf{T}}$.

Such a Lagrange basis allows us to formally introduce error bounds for the VSK interpolant that depend on the *similarities* between ψ and f.

Proposition 16.1.1 *Let ψ be a scaling function, s_{ψ}^{Ψ} be the VSK interpolant of the function ψ and let $\kappa^{\Psi}:\Omega\times\Omega\longrightarrow\mathbb{R}$ be the VSK corresponding to the strictly positive definite and symmetric kernel κ. Let $s^{\Psi}:\Omega\longrightarrow\mathbb{R}$ be the VSK interpolant. Then, the following point-wise error bound holds true*

$$\left|\left(f-s^{\Psi}\right)(\boldsymbol{x})\right| \le \left|\left(f-s_{\psi}^{\Psi}\right)(\boldsymbol{x})\right| + \|\boldsymbol{f}-\boldsymbol{\psi}\|_{\infty}\lambda^{\Psi}(\boldsymbol{x}), \tag{16.1}$$

where

$$\lambda^{\Psi}(\boldsymbol{x}) = \sum_{j=1}^{N} |u_j^{\Psi}(\boldsymbol{x})|$$

is the Lebesgue function for the VSK interpolant, and $\boldsymbol{f}-\boldsymbol{\psi} = (f(\boldsymbol{x}_1)-\psi(\boldsymbol{x}_1),\dots,f(\boldsymbol{x}_N)-\psi(\boldsymbol{x}_N))^{\mathsf{T}} = (f_1-\psi_1,\dots,f_N-\psi_N)^{\mathsf{T}}$.

Proof. We trivially have

$$\begin{aligned}\left|\left(f-s^{\Psi}\right)(\boldsymbol{x})\right| &= \left|\left(f-s_{\psi}^{\Psi}\right)(\boldsymbol{x})+\left(s_{\psi}^{\Psi}-s^{\Psi}\right)(\boldsymbol{x})\right| \\ &\leq \left|\left(f-s_{\psi}^{\Psi}\right)(\boldsymbol{x})\right|+\left|\left(s^{\Psi}-s_{\psi}^{\Psi}\right)(\boldsymbol{x})\right|.\end{aligned}$$

To bound the second term in the previous inequality, we observe that

$$\left|\left(s^{\Psi}-s_{\psi}^{\Psi}\right)(\boldsymbol{x})\right| = \left|s_{f-\psi}^{\Psi}(\boldsymbol{x})\right| = \sum_{j=1}^{N}\left|(f_j-\psi_j)u_j^{\Psi}(\boldsymbol{x})\right| \leq \|\boldsymbol{f}-\boldsymbol{\psi}\|_{\infty}\lambda^{\Psi}(\boldsymbol{x}),$$

and the thesis follows. □

Equivalently to (16.1), we might write

$$\left|\left(f-s^{\Psi}\right)(\boldsymbol{x})\right| \leq \left|\left(f-s_{\psi}^{\Psi}\right)(\boldsymbol{x})\right| + \|\boldsymbol{f}-\boldsymbol{\psi}\|_{\infty}\Lambda^{\Psi},$$

where Λ^{Ψ} is the Lebesgue constant given by

$$\Lambda^{\Psi} = \sup_{\boldsymbol{x}\in\mathbb{R}^d}\sum_{j=1}^{N}|u_j^{\Psi}(\boldsymbol{x})|.$$

We point out that to eliminate the second term on the right-hand side of (16.1), we only need to set the nodal values of ψ equal to those of f; this could be trivially achieved since the interpolation conditions are known. Nevertheless, the first term suggests that the scaling function ψ should be close to f on the whole domain Ω, since in this way s_{ψ}^{Ψ} should approach s^{Ψ} and f in turn, following classical error bounds based either on the power function or on the fill-distance (see Wendland, 2005, Theorem 11.4 and Theorem 11.13). Moreover, we stress that taking $f \equiv \psi$ is an unrealistic choice, since f is unknown, and we have no theoretical or numerical evidence that $|(f-s^f)(\boldsymbol{x})| \leq |(f-s^{\Psi})(\boldsymbol{x})|$ for each ψ and $\boldsymbol{x}\in\Omega$.

Finally, regarding error bounds in native spaces, we have Corollary 16.1.1.

Corollary 16.1.1 *Let $\kappa^{\Psi}:\Omega\times\Omega\longrightarrow\mathbb{R}$ be the VSK corresponding to the strictly positive definite and symmetric kernel κ and $f\in\mathscr{N}_{\kappa^{\Psi}}(\Omega)$, then*

$$\begin{aligned}\left|\left(f-s^{\Psi}\right)(\boldsymbol{x})\right| \leq{} & \left(\|f\|_{\mathscr{N}_{\kappa^{\Psi}}(\Omega)} + \boldsymbol{\psi}^{\mathsf{T}}(\mathsf{K}^{\Psi})^{-1}\boldsymbol{\psi}\right)\|\kappa^{\Psi}\|_{\mathscr{N}_{\kappa^{\Psi}}(\Omega)} \\ & + \|\boldsymbol{f}-\boldsymbol{\psi}\|_{\infty}\lambda^{\Psi}(\boldsymbol{x}).\end{aligned}$$

Proof. In Bozzini et al. (2015, Theorem 2), the authors proved that the native spaces $\mathscr{N}_{\kappa}(\Omega)$ and $\mathscr{N}_{\kappa^{\Psi}}(\Omega)$ are isometric. Then, by the Cauchy–Schwarz inequality and the reproducing property, we have

$$\begin{aligned}\left|\left(f-s_{\psi}^{\Psi}\right)(\boldsymbol{x})\right| &= \left(f-s_{\psi}^{\Psi}, \kappa^{\Psi}(\cdot,\boldsymbol{x})\right)_{\mathscr{N}_{\kappa^{\Psi}}(\Omega)} \\ &\leq \|f-s_{\psi}^{\Psi}\|_{\mathscr{N}_{\kappa^{\Psi}}(\Omega)} \|\kappa^{\Psi}\|_{\mathscr{N}_{\kappa^{\Psi}}(\Omega)} \\ &\leq \left(\|f\|_{\mathscr{N}_{\kappa^{\Psi}}(\Omega)} + \|s_{\psi}^{\Psi}\|_{\mathscr{N}_{\kappa^{\Psi}}(\Omega)}\right) \|\kappa^{\Psi}\|_{\mathscr{N}_{\kappa^{\Psi}}(\Omega)},\end{aligned}$$

and as $\|s_{\psi}^{\Psi}\|_{\mathscr{N}_{\kappa^{\Psi}}(\Omega)} = \boldsymbol{\psi}^{\mathsf{T}}(\mathsf{K}^{\Psi})^{-1}\boldsymbol{\psi}$, the thesis follows. □

Finally, we now have some theoretical evidence that the function ψ should approximate f. However, it is worth clarifying that no optimal properties have been proved and that, since the function f is unknown in practice, we do not have any criteria or algorithms to define the *best* scaling function automatically. Hence, we implement an NN that confirms our findings and returns a safe (i.e., nearly optimal) scaling function ψ. More precisely, numerically, we will observe that for the practical implementation of the δNN–VSKs, the demanding pointwise convergence of the function ψ to f can typically be relaxed, as reproducing some of the key features of the target usually leads to accurate approximations. Further comments are provided in Section 16.2.3.

16.2 Learning VSKs via Discontinuous Neural Networks

16.2.1 δNN–VSKs: Learning Nearly Optimal Scaling Functions

The NN that we consider is called δNN, and its first usage (Della Santa and Pieraccini, 2023) was devoted to detecting discontinuity interfaces while approximating discontinuous functions. This type of NN is characterized by the introduction of learnable discontinuities in the layers, legitimized by the universal approximation theorems (Kidger and Lyons, 2010), and hence might return a discontinuous output. Before introducing the discontinuous layers that characterize δNNs, we recall that a standard FC layer, from n_{in} units to n_{out} units, can be described by the function

$$\mathscr{L}(\boldsymbol{x}) = \boldsymbol{\sigma}\left(\mathsf{W}^{\mathsf{T}}\boldsymbol{x}+\boldsymbol{b}\right), \quad \boldsymbol{x} \in \mathbb{R}^{n_{\text{in}}}, \tag{16.2}$$

where $\mathsf{W} \in \mathbb{R}^{n_{\text{in}} \times n_{\text{out}}}$ is the weight matrix, $\boldsymbol{b} \in \mathbb{R}^{n_{\text{out}}}$ is the bias vector, and $\boldsymbol{\sigma}$ is the element-wise application of the activation function $\sigma : \mathbb{R} \to \mathbb{R}$.

The characterizing function of a discontinuous layer is a function $\delta\mathscr{L}$: $\mathbb{R}^{n_{\text{in}}} \to \mathbb{R}^{n_{\text{out}}}$ such that

$$\mathscr{L}(\boldsymbol{x}) = \boldsymbol{\sigma}\left(\mathsf{W}^{\mathsf{T}}\boldsymbol{x}+\boldsymbol{b}\right) + \boldsymbol{\alpha} \odot \mathscr{H}\left(\mathsf{W}^{\mathsf{T}}\boldsymbol{x}+\boldsymbol{b}\right), \quad \boldsymbol{x} \in \mathbb{R}^{n_{\text{in}}}, \tag{16.3}$$

where $\mathscr{H}$ is the element-wise application of the Heaviside function and $\boldsymbol{\alpha} \in \mathbb{R}^{n_{\text{out}}}$ is the vector of *trainable discontinuity jumps*. In brief, for each $j = 1, \ldots, n_{\text{out}}$, (16.3) is equivalent to adding to the j-th element of (16.2) a jump of height α_j if the j-th row of $\mathsf{W}^\mathsf{T}\boldsymbol{x} + \boldsymbol{b}$ is nonnegative (otherwise, no jump is applied).

The main property of an NN embedded with discontinuous layers (i.e., a δNN) is the ability to return both an approximation of the target function and an approximation of its discontinuity interfaces. Specifically, in a δNN, the weight matrices and the bias vectors of the discontinuous layers have the role of learning the discontinuity interfaces. In contrast, the discontinuity jump parameters serve to learn the jump height at the discontinuity interfaces.

The reason why we are interested in δNNs to define the scaling function ψ lies in the fact that, in principle, ψ can be discontinuous. The novelty in this case is that we do not learn the discontinuities of f; we learn the function ψ directly without imposing any constraints on its discontinuities.

We define the δNN as a parametric scaling function for the VSKs, aiming to find the *optimal* one for the interpolant of f; that is, we train a δNN $\psi_{\boldsymbol{\theta}} : \mathbb{R}^d \to \mathbb{R}^m$, with trainable parameters denoted by $\boldsymbol{\theta} \in \mathbb{R}^p$, and adapt the interpolation coefficients $\boldsymbol{c} \in \mathbb{R}^N$ by minimizing the interpolation error of the VSK interpolant.

More precisely, we build an NN model $\mathscr{M}_{\boldsymbol{\theta},\boldsymbol{\eta}} : \mathbb{R}^d \to \mathbb{R}^m \times \mathbb{R}^N$ that pairs the δNN $\psi_{\boldsymbol{\theta}}$ and a model $\boldsymbol{c}_\eta := \boldsymbol{c}(\boldsymbol{\eta})$ for the coefficients, with $\boldsymbol{\theta} \in \mathbb{R}^p, \boldsymbol{\eta} \in \mathbb{R}^q$ trainable parameters, such that

$$\mathscr{M}_{\boldsymbol{\theta},\boldsymbol{\eta}}(\boldsymbol{x}) = (\psi_{\boldsymbol{\theta}}(\boldsymbol{x}), \boldsymbol{c}_\eta),$$

for each $\boldsymbol{x} \in \mathbb{R}^d$.

Training of $\mathscr{M}_{\boldsymbol{\theta},\boldsymbol{\eta}}$ is obtained by minimizing the loss

$$\begin{aligned} \ell_{X_N}(\boldsymbol{\theta}, \boldsymbol{\eta}) &:= \frac{1}{N} \| \boldsymbol{f} - \mathsf{K}^{\Psi_{\boldsymbol{\theta}}} \boldsymbol{c}_{\boldsymbol{\eta}} \|_2^2 \\ &= \frac{1}{N} \sum_{i=1}^{N} \left(f_i - \sum_{k=1}^{N} c_k(\boldsymbol{\eta})\ \kappa((\boldsymbol{x}_i, \psi_{\boldsymbol{\theta}}(\boldsymbol{x}_i)) \ , \ (\boldsymbol{x}_k, \psi_{\boldsymbol{\theta}}(\boldsymbol{x}_k))) \right)^2 , \end{aligned} \tag{16.4}$$

where a global minimum $(\boldsymbol{\theta}^*, \boldsymbol{\eta}^*) \in \mathbb{R}^p \times \mathbb{R}^q$ for (16.4) is such that $\ell_{X_N}(\boldsymbol{\theta}^*, \boldsymbol{\eta}^*) = 0$ (i.e., interpolation satisfied).

We drive the reader's attention toward the fact that (16.4) is defined using the whole training data $\{(\boldsymbol{x}_i, f_i)\}_{i=1}^N$ and not using random mini-batches sampled from them; indeed, since the main task is the interpolation, we prefer a deterministic training procedure instead of a stochastic one.

Remark 16.2.1 *Let* $(\boldsymbol{\theta}^{\text{fin}}, \boldsymbol{\eta}^{\text{fin}})$ *be the parameters obtained at the end of the* *training by minimizing* (16.4). *The interpolation conditions corresponding to* $(\boldsymbol{\theta}^{\text{fin}}, \boldsymbol{\eta}^{\text{fin}})$ *are likely to be only approximately satisfied; indeed, stopping criteria are adopted, and it is possible that training stops before reaching a given tolerance. Therefore, it is recommended to build a model based on* $\boldsymbol{\theta}^{\text{fin}}$ *and an updated version of* $\boldsymbol{c}$*, namely* $\boldsymbol{c} = (\mathsf{K}^{\Psi_{\boldsymbol{\theta}^{\text{fin}}}})^{-1}\boldsymbol{f}$. *In case an iterative solver is used to solve the linear system,* $\boldsymbol{c}_{\boldsymbol{\eta}^{\text{fin}}}$ *can be used as a starting guess.*

Summarizing the procedure, the δNN–VSK interpolant for a target function f is constructed as in (16.4) where the coefficients $\boldsymbol{c} = (c_1, \ldots, c_N)^\mathsf{T}$ are calculated as reported in the Remark 16.2.1.

The numerical experiments show that the scaling function learned via the δNN–VSK method is, indeed, a *nearly optimal choice* and that it *closely mimics the target's shape*.

In Section 16.2.2, we propose an alternative way to compute the VSK interpolant, namely the *VSK-f approach* , which directly learns the target and uses this approximation as a scaling function.

16.2.2 VSK-f: Using an Approximation of f

Another possible approach to defining a proper scaling function is based on the claim stated at the end of Section 16.1. Specifically, Proposition 16.1.1 suggests that a good approximation of the target can be used as a "safe" scaling function for a VSK interpolant. Therefore, we can renounce to find a nearly optimal scaling function minimizing a loss like (16.4), preferring to train a δNN for directly approximating f, instead. However, we recall that the amount of data typically used for interpolation (i.e., N) is often insufficient for an acceptable approximation of f by δNN models; therefore, finding a VSK interpolant of f remains necessary.

In detail, given a δNN $\psi_{\boldsymbol{\theta}}$, we train it to approximate f, using the interpolation data as the training set and through a classic stochastic training procedure. Therefore, the trained model $\psi_{\boldsymbol{\theta}^{\text{fin}}}$ will be an approximation of f, and the larger N, the finer the approximation quality.

Remark 16.2.2 *The main advantage of using VSKs-f with respect to δNN–VSKs is that training* $\psi_{\boldsymbol{\theta}}$ *for approximating* f *is easier than training the model* $\mathscr{M}_{\boldsymbol{\theta},\boldsymbol{\eta}}$ *that simultaneously learns the coefficients of the approximant and the scaling function. Of course, the price of an easier procedure is the risk of using a safe scaling function that is not tailored to the considered kernel basis.*

16.2.3 Results

We test the proposed procedure on synthetic datasets obtained by sampling three selected functions at given sets of quasi-uniform nodes.

We take as test functions the well-known *Franke* function f_1, which is continuous, being the sum of exponential terms, and the discontinuous test functions:

$$f_2(x_1,x_2) = \begin{cases} -\mathrm{e}^{(-(x_1-0.5)^2+(x_2-0.5)^2)}, & \text{if } \ ||\boldsymbol{x}-(0.5,0.5)||_2^2 \geq 0.08 \\ \sin(x_1)+4\sin(x_2), & \text{otherwise.} \end{cases}$$

We sample these functions at $N = [27^2, 33^2, 39^2]$ scattered Halton data in $[0,1]^2$ and interpolate them with respect to

- classical RBF,
- δNN–VSKs (see Section 16.2.1);
- VSKs-f (see Section 16.2.2).

As accuracy indicators, we report in Table 16.1 the MAE, SSIM, and the mean-squared error (MSE):

$$\text{MSE} = \frac{1}{M}\sum_{k=1}^{M}(f(\boldsymbol{z}_k) - s^{\Psi}(\boldsymbol{z}_k))^2,$$

where $\boldsymbol{z}_1, \ldots, \boldsymbol{z}_M$ are M evaluation data in Ω.

The kernel used to approximate the function f_1 (independently of the method used) is the Gaussian C^∞ function, while for f_2 we use the Matérn C^2 kernel.

All results are reproducible, as the Python code is freely available at

`https://github.com/Fra0013To/VSKlearning/tree/paper2024.`

In Figures 16.1–16.4, we depict all the surfaces obtained. In particular, we plot both their top view, which is essential for clearly seeing the improvements in the discontinuous case, and their 3D counterpart, false-colored with the absolute error.

In all the test cases considered, we observe (see Table 16.1) that the δNN–VSK and VSK-f methods behave better than classical kernels. Precisely, for the continuous test function f_1, the standard RBFs already give good approximations. At the same time, in the discontinuous cases the δNN–VSK and VSK-f methods provide a consistent improvement on the approximation accuracy (see Table 16.1 and Figures 16.1–16.4); moreover, in the latter case, the improvement in terms of SSIM (refer to Table 16.1) of the VSKs-based methods is rather impressive. Such results are because when data are smooth, VSKs might only improve already quite good approximation scores returned by usual kernels. On the other hand, in discontinuous cases, the most challenging task is identifying and reconstructing the faults; therefore, any discontinuous scaling function that shares the same discontinuities as the target provides a positive contribution.

Table 16.1 MAEs, MSEs, and SSIMs for the test functions f_1 and f_2

		MAE			MSE			SSIM		
	n	RBFs	δNN-VSKs	VSKs-f	RBFs	δNN-VSKs	VSKs-f	RBFs	δNN-VSKs	VSKs-f
f_1	729	4.99e-2	8.98e-3	3.57e-3	4.24e-3	1.42e-4	2.43e-5	0.8712	0.9820	0.9943
	1089	2.38e-2	3.48e-3	2.11e-3	9.88e-4	2.33e-5	8.25e-6	0.9533	0.9969	0.9980
	1521	3.60e-4	3.60e-4	2.47e-4	3.84e-7	3.83e-7	2.70e-7	0.9997	0.9997	0.9999
f_2	729	1.97e-1	2.09e-2	1.65e-2	5.76e-2	3.72e-3	3.84e-3	0.5883	0.9074	0.9253
	1089	1.45e -1	8.06e-3	1.29e-2	3.37e-2	2.12e-3	2.18e-3	0.5885	0.9556	0.9486
	1521	6.65e-2	5.68e-3	8.37e-3	1.33e-2	1.89e-3	2.37e-3	0.6453	0.9696	0.9559

In all cases, we recall that no information is provided on the target's continuity or discontinuity, nor on the selection of the scaling function. Therefore, the continuous case is a compelling test case, as it demonstrates the flexibility of the δNN architecture for learning continuous targets. In other words, if one does not have information on the smoothness of the data, then the δNN–VSK approach is a practical and robust choice for approximating the target.

Moreover, regarding the scaling function learned via δNNs, we observe that for both the continuous test f_1 and the discontinuous one f_2, the δNN–VSK method learns nearly optimal scaling functions that mimic the targets, while the VSK-f method supports the theoretical claims of safety in selecting a scaling function that not only mimics but approximates the target; refer to (e) and (f) of Figures 16.1 and 16.2.

Summarizing, from the analysis of the interpolation results, we observe the following:

- Our VSKs-based methods provide reliable approximations both for the continuous case and for discontinuous test functions (see Table 16.1 and Figures 16.3–16.4).
- The δNN–VSK method shows that it is able to learn a nearly optimal scaling function automatically, and as a numerical confirmation of our claims, the identified scaling functions $\bar{f}_{\boldsymbol{\theta}^{\text{fin}}}$ are such that they *mimic* the target f. This mimicking behavior can be detailed (see, e.g., Figure 16.1) or "generic" (see, e.g., Figure 16.2).
- The results obtained via the VSK-f method support our theoretical claim reported at the end of Section 16.1; that is, using a scaling function $\bar{f}_{\boldsymbol{\theta}^{\text{fin}}} \approx f$ is a safe choice that guarantees a good interpolation accuracy, even if not necessarily the *optimal one* (see Table 16.1).

Concerning the second item, it is worth remarking that the function $\bar{f} :=$

$\bar{f}_{\boldsymbol{\theta}^{\text{fin}}}$ of the δNN–VSK method is such that $\bar{f} \approx \gamma f$, where $\gamma : \mathbb{R}^d \longrightarrow \mathbb{R}$ is a continuous function. This is consistent with our implementation of the δNN and its loss (16.4). Let us write the VSKs with their explicit dependence on the shape parameter and on the radial distance, that is, for $\boldsymbol{x}, \boldsymbol{z} \in \Omega$:

$$\kappa^{\gamma f}(\boldsymbol{x},\boldsymbol{z}) = \phi\left(\varepsilon\sqrt{\sum_{k=1}^{d}(x^{(k)} - z^{(k)})^2 + ((\gamma f)(\boldsymbol{x}) - (\gamma f)(\boldsymbol{z}))^2}\right)$$

$$= \phi\left(\sqrt{\sum_{k=1}^{d}\varepsilon^2\left(x^{(k)} - z^{(k)}\right)^2 + \varepsilon^2\left(\gamma(\boldsymbol{x})f(\boldsymbol{x}) - \gamma(\boldsymbol{z})f(\boldsymbol{z})\right)^2}\right).$$

Then, the fact that the function $\bar{f}$ mimics the shape of the function f but sometimes possibly out of scale is a consequence of the fact that the δNN itself tries to *adjust*, via the function γ, the fixed shape parameter ε. However, the most relevant result of these experiments is that we always observe some similarities between the targets and the recovered scaling functions (consistent with Proposition 16.1.1) and that the proposed VSK-based methods correctly identify nearly optimal/safe scaling functions.

In this section, by taking advantage of the cardinal form of the VSK interpolant, we introduce new error bounds that relate the accuracy of the VSK interpolant to the definition of the associated scaling function. These bounds indicate that the closer the scaling function is to the target, the smaller the pointwise error. We further propose a way to learn the scaling function using δNNs, and this data-driven approach offers a user-independent solution that enhances the flexibility of VSKs for meshfree approximation tasks and numerically shows that the learned scaling function captures the key features of the target, as expected.

The numerical tests show that our two possible implementations of the VSKs, namely the δNN–VSKs and the VSKs-f, behave better than the classical kernel interpolant. The main advantage of the VSKs-f is its ease of implementation, as it directly learns the target, which is then used as a scaling function. Nevertheless, the justification for using such a scheme is subordinated to the theoretical results and to the fact that δNN–VSKs, which can learn the nearly optimal scaling function without any constraints, return scaling functions that truly resemble the targets.

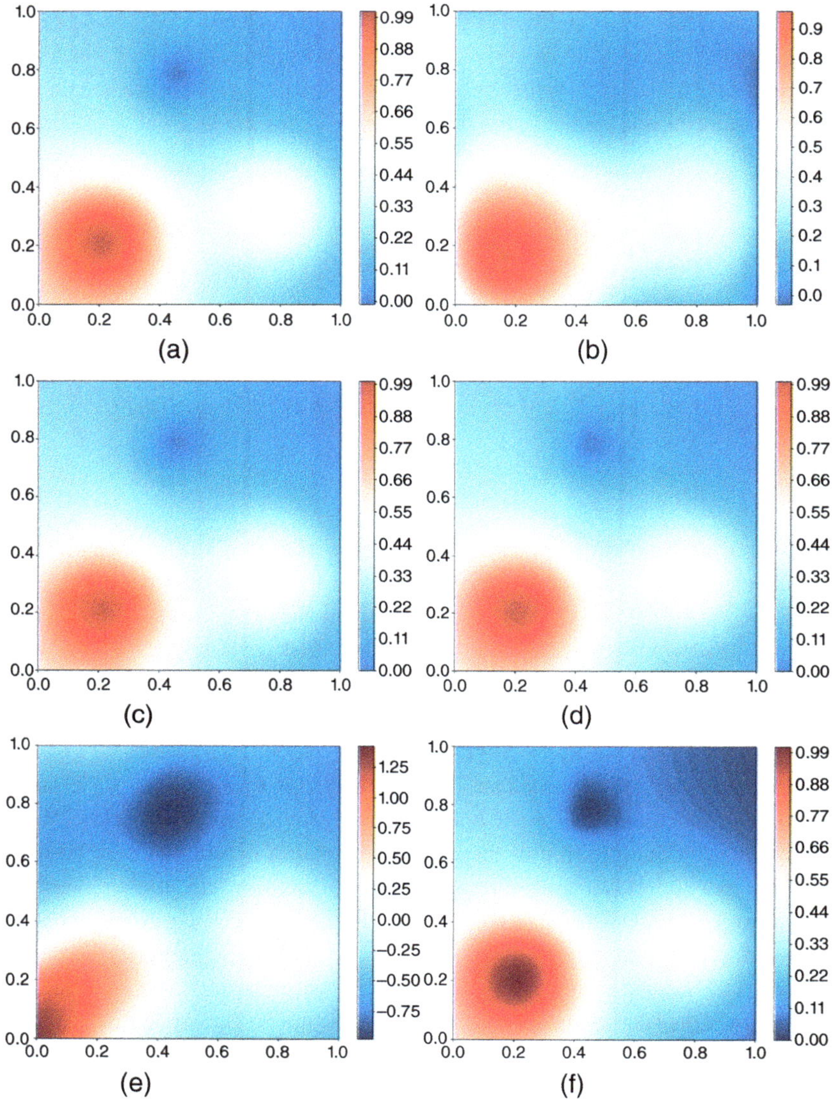

Figure 16.1 Interpolation results for f_1. Case $n = 33^2 = 1089$ interpolation points, Gaussian Kernel. Shared color bars for (a) target; (b) RBFs; (c) δNN–VSKs; (d) VSKs-f; custom color bars for the scaling functions (subfigures (e) δNN–VSKs scaling func.; (f) VSKs-f, scaling func.).

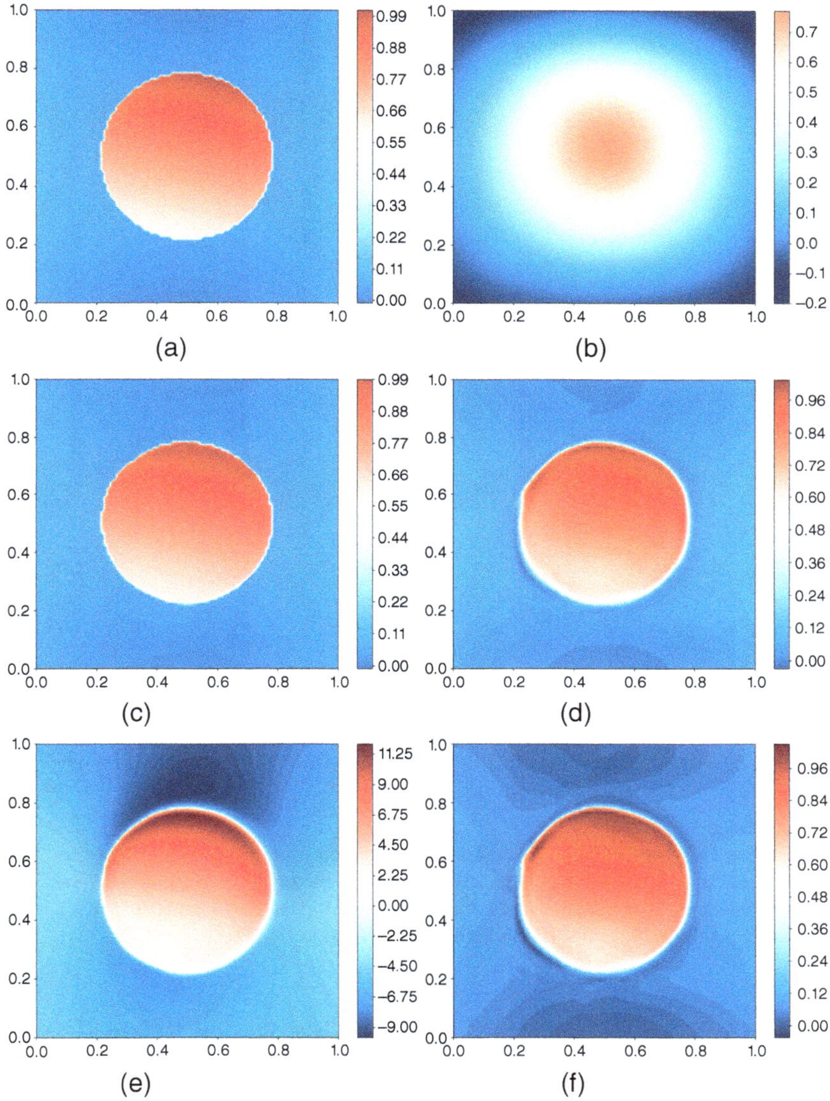

Figure 16.2 Interpolation results for f_2. Case $n = 33^2 = 1089$ interpolation points, Matérn-C^2 Kernel. Shared color bars for (a) target; (b) RBFs; (c) δNN–VSKs; (d) VSKs-f; custom color bars for the scaling functions ((e) δNN–VSKs scaling func.; (f) VSKs-f, scaling func.)

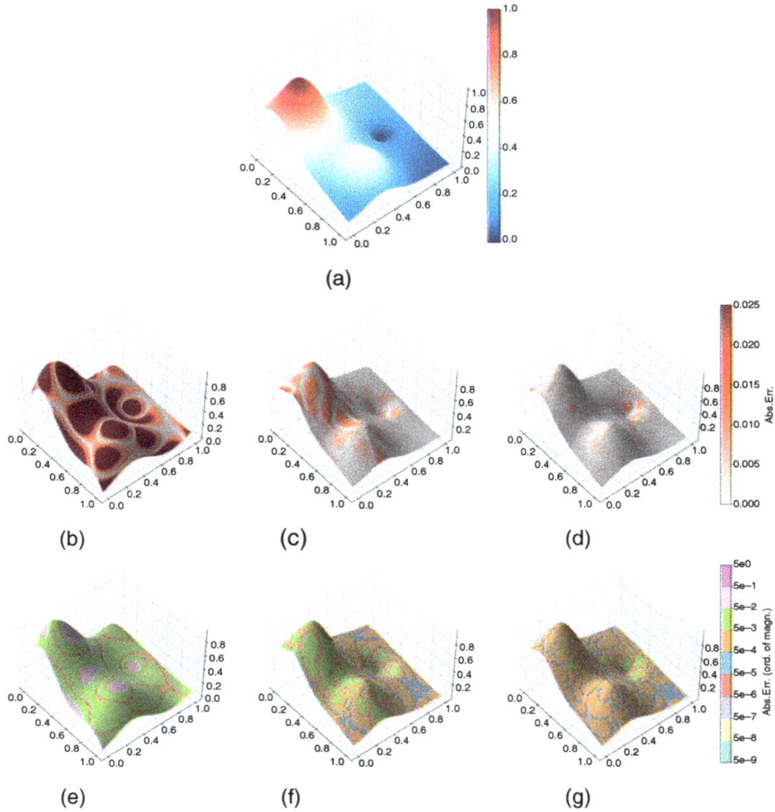

Figure 16.3 Interpolation results for f_1 (a), colored w.r.t. interpolation error (absolute error). In (e), (f), and (g), colors describe the order of magnitude of the absolute error. Shared colorscales for (b) RBFs; (c) δNN-VSKs; and (d) VSKs-f and for (e) RBFs; (f) δNN-VSKs; and (g) VSKs-f.

16.3 An Extension of the Variably Scaled Setting

In this section, we first present a generalization of VSKs beyond real-valued scaling functions. Then, we outline some choosing strategies for the scaling function for regression and classification tasks, presenting results from Campi et al. (2021).

Referring to Definition 9.1.1, we do not restrict ourselves to real-valued scaling functions here. Indeed, the first step is to consider the multidimensional map $\psi : \mathbb{R}^d \longrightarrow \Lambda$, $\Lambda \subset \mathbb{R}^m$, $d, m \geq 1$, and the function $\Psi : \mathbb{R}^d \longrightarrow \mathbb{R}^{d+m}$ defined as

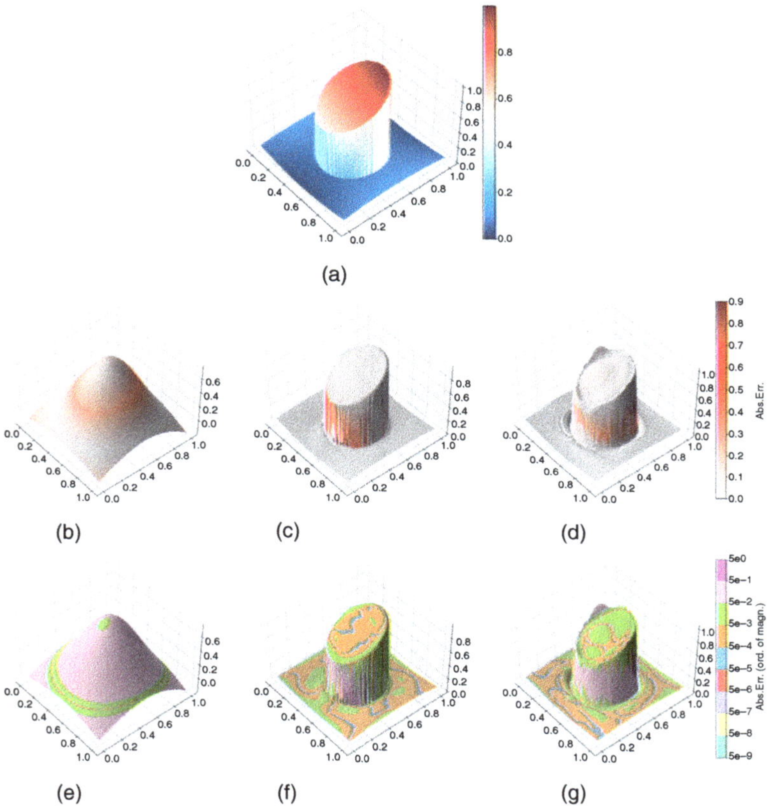

Figure 16.4 Interpolation results for f_2 (see Figure 16.2), colored w.r.t. interpolation error (absolute error). In (e), (f), and (g), colors describe the order of magnitude of the absolute error. Shared color bars for (b) RBFs; (c) δNN-VSKs; and (d) VSKs-f and for (e) RBFs; (f) δNN-VSKs; and (g) VSKs-f.

$$\Psi(\boldsymbol{x}) := (\boldsymbol{x}, \psi(\boldsymbol{x})).$$

where $\boldsymbol{x} \in \mathbb{R}^d$.

Then, let $\widetilde{\boldsymbol{x}}, \widetilde{\boldsymbol{y}} \in \mathbb{R}^{d+m}$ and let κ be a positive definite kernel on $\mathbb{R}^{(d+m)\times(d+m)}$. As presented in Section 12.2, we can write

$$\kappa(\widetilde{\boldsymbol{x}}, \widetilde{\boldsymbol{y}}) = \langle \hbar(\widetilde{\boldsymbol{x}}), \hbar(\widetilde{\boldsymbol{y}}) \rangle_{\mathscr{H}},$$

where $\mathscr{H}$ and $\hbar : \mathbb{R}^{d+m} \longrightarrow \mathscr{H}$ are a feature space and a feature map related to κ, respectively. We can define on $\mathbb{R}^{d\times d}$ the VSK κ^{Ψ} as

$$\kappa^{\Psi}(\boldsymbol{x}, \boldsymbol{y}) = \kappa(\Psi(\boldsymbol{x}), \Psi(\boldsymbol{y})).$$

Remark 16.3.1 *The scaling function* ψ *can be interpreted as a map that extracts* m *features from the available data points. Such new features are then juxtaposed with the originals in the construction of the VSK, by means of the function* Ψ.

Remark 16.3.2 *We adopt the term VSK even if discontinuous scaling maps are taken into account. Indeed, in many situations, it is neither enjoyable nor meaningful to focus on discontinuity properties, for example, when dealing with integer-valued discrete data.*

As a consequence of Proposition 12.2.1 (v) in Section 12.2.2, the kernel κ^{Ψ} is positive definite. Thus, it is possible to write

$$\kappa^{\Psi}(\boldsymbol{x},\boldsymbol{y}) = \langle \hbar^{\Psi}(\boldsymbol{x}), \hbar^{\Psi}(\boldsymbol{y})\rangle_{\mathscr{H}^{\Psi}}, \quad \boldsymbol{x},\boldsymbol{y} \in \Omega.$$

where $\mathscr{H}^{\Psi}$ and $\hbar^{\Psi}: \mathbb{R}^d \longrightarrow \mathscr{H}$ are a feature space and feature map related to κ^{Ψ}, respectively.

For the analysis of the variably scaled setting, we take $\Omega \subset \mathbb{R}^d$ and $X = \{\boldsymbol{x}_k,\ k = 1,\ldots,N\} \subset \Omega$ as a set of data points. Moreover, we focus on linear and Gaussian kernels defined in (12.3) and (12.5), respectively, since they are widely used in applications, for example, with SVMs.

16.3.1 On the Linear Kernel

Let κ_L^{Ψ} be the linear VSK. Its kernel matrix K^{Ψ} can be constructed as follows.

Theorem 16.3.1 *Let* $X = \{\boldsymbol{x}_k,\ k = 1,\ldots,N\} \subset \Omega$ *and let* κ_L^{Ψ} *denote the linear kernel on* $\Omega \times \Omega$ *and with* K *the related kernel matrix on* X*. Then, the VSK matrix* K^{Ψ} *on* X *related to* κ_L^{Ψ} *on* $\Omega \times \Omega$ *is given by*

$$\mathsf{K}^{\Psi} = \mathsf{K} + \mathsf{K}^{\psi},$$

where $\mathsf{K}^{\psi}_{ij} = \psi(\boldsymbol{x}_i) \cdot \psi(\boldsymbol{x}_j)$, $i,j = 1,\ldots,N$.

Proof. For $\boldsymbol{x},\boldsymbol{y} \in \Omega$ we have that:

$$\kappa_L^{\Psi}(\boldsymbol{x},\boldsymbol{y}) = (\boldsymbol{x},\psi(\boldsymbol{x})) \cdot (\boldsymbol{y},\psi(\boldsymbol{y})) = \boldsymbol{x}\cdot\boldsymbol{y} + \psi(\boldsymbol{x})\cdot\psi(\boldsymbol{y}),$$

thus

$$\mathsf{K}^{\Psi} = \mathsf{K} + \mathsf{K}^{\psi}.$$

□

16.3.1.1 SR of the Linear VSK

Concerning the SR (see Definition 12.2.1) of κ_L^{Ψ}, we have Proposition 16.3.1

Proposition 16.3.1 *Let* $X = \{\boldsymbol{x}_k,\ k = 1,\dots,N\} \subset \Omega$ *and let* K^{Ψ}, K, *and* K^{ψ} *be as in Theorem 16.3.1. If* K *and* K^{ψ} *are nonnegative, that is, so that all entries are nonnegative, then*

$$\frac{tr(\mathsf{K})}{tr(\mathsf{K}^{\Psi})} \le \frac{SR(\mathsf{K})}{SR(\mathsf{K}^{\Psi})} \le \frac{\|\mathsf{K}^{\Psi}\|_F}{\|\mathsf{K}\|_F}.$$

Proof. Under our assumptions, we have that

$$\frac{\mathrm{tr}(\mathsf{K})}{\mathrm{tr}(\mathsf{K}^{\Psi})} \le 1,$$

and

$$\frac{\|\mathsf{K}^{\Psi}\|_F}{\|\mathsf{K}\|_F} \ge 1.$$

Then, the thesis follows by taking into account the definition of SR. □

16.3.1.2 Spectrum of the Linear VSK

Since they are Gramiam matrices, K^{Ψ} and K^{ψ} are positive semi-definite. Concerning the spectrum of the VSK matrix K^{Ψ}, by virtue of Weyl's inequality (see, e.g., Bhatia, 1997, section III.2, p. 62), we have that

$$\lambda_{\min}(\mathsf{K}) \le \lambda_{\min}(\mathsf{K} + \mathsf{K}^{\psi}) = \lambda_{\min}(\mathsf{K}^{\Psi}).$$

16.3.2 On the Gaussian Kernel

Here, we present some results related to the kernel matrix of the Gaussian VSK κ_G^{Ψ} in the variably scaled setting.

Theorem 16.3.2 *Let* $X = \{\boldsymbol{x}_k,\ k = 1,\dots,N\} \subset \Omega$ *and let* κ_G^{Ψ} *denote the Gaussian kernel on* $\Omega \times \Omega$ *and with* K *the related kernel matrix on* X*. Then, the VSK matrix* K^{Ψ} *on* X *related to* κ_G^{Ψ} *on* $\Omega \times \Omega$ *is given by*

$$\mathsf{K}^{\Psi} = \mathsf{K} \circ \mathsf{K}^{\psi},$$

where $\mathsf{K}_{ij}^{\psi} = e^{-\|\psi(\boldsymbol{x}_i) - \psi(\boldsymbol{x}_j)\|^2}$, $i, j = 1,\dots,N$, K *the standard kernel matrix and* $\circ$ *denotes the Hadamard matrix product.*

Proof. For $\boldsymbol{x}, \boldsymbol{y} \in \Omega$, we can write

$$\kappa^{\Psi}(\boldsymbol{x},\boldsymbol{y}) = e^{-\|\boldsymbol{x}-\boldsymbol{y}\|^2 + \|\psi(\boldsymbol{x}) - \psi(\boldsymbol{y})\|^2} = e^{-\|\boldsymbol{x}-\boldsymbol{y}\|^2}\, e^{-\|\psi(\boldsymbol{x}) - \psi(\boldsymbol{y})\|^2}.$$

Therefore, we obtain

$$\mathsf{K}_{ij}^{\Psi} = e^{-\|\boldsymbol{x}_i - \boldsymbol{x}_j\|^2}\, e^{-\|\psi(\boldsymbol{x}_i) - \psi(\boldsymbol{x}_j)\|^2}.$$

In other words,

$$\mathsf{K}^{\Psi} = \mathsf{K} \circ \mathsf{K}^{\psi}.$$

□

Concerning the Hadamard product, we report here a useful result that dates back to 1911 in Schur (1911). We also refer to Karoui (2010, Lemma A.5) and Horn and Zhang (2010, Lemma 2.1).

Theorem 16.3.3 *If* E *and* $\mathsf{M} \in \mathbb{R}^{N \times N}$ *are positive definite matrices, denoting* $\lambda_{\min}$ *and* $\lambda_{\max}$ *the smallest and largest eigenvalue of a matrix, we have that*

$$\lambda_{\min}(\mathsf{E}) \min_{i=1,\dots,N} \mathsf{M}_{ii} \leq \lambda_i(\mathsf{E} \circ \mathsf{M}) \leq \lambda_{\max}(\mathsf{E}) \max_{i=1,\dots,N} \mathsf{M}_{ii}.$$

16.3.2.1 SR of the Gaussian VSK

Here, we investigate the SR of the Gaussian VSK. To this aim, we first provide upper and lower bounds for the Frobenius norm of the kernel matrix K in terms of the Frobenius norm of K^{Ψ}.

Theorem 16.3.4 *Let* $X = \{\boldsymbol{x}_k,\ k = 1, \dots, N\} \subset \Omega$ *and let* K^{Ψ}, K *and* K^{ψ} *be as in Theorem 16.3.2. We have*

$$\|\mathsf{K}^{\Psi}\|_F \leq \|\mathsf{K}\|_F \leq \|\mathsf{K}\|_F \|\mathsf{K}^{\psi}\|_F.$$

Proof. The univariate function φ_G associated with κ_G is non-increasing. Then, we can write

$$\varphi_G\left(\|\boldsymbol{x}-\boldsymbol{y}\|^2\right) \geq \varphi_G\left(\|\boldsymbol{x}-\boldsymbol{y}\|^2 + \|\psi(\boldsymbol{x}) - \psi(\boldsymbol{y})\|^2\right),$$

which implies

$$\mathsf{K}_{ij} \geq \mathsf{K}^{\Psi}_{ij} \geq 0, \quad i, j = 1, \dots, N.$$

Therefore,

$$\|\mathsf{K}\|_F \geq \|\mathsf{K}^{\Psi}\|_F.$$

Moreover, since $\varphi_G(0) = 1$, we have that

$$\|\mathsf{K}^{\psi}\|_F \geq \sqrt{\sum_{i=1}^{N} \left(\mathsf{K}^{\psi}_{ii}\right)^2} = \sqrt{N(\varphi_G(0))^2} \geq 1,$$

and then

$$\|\mathsf{K}^{\Psi}\|_F \leq \|\mathsf{K}\|_F \leq \|\mathsf{K}\|_F \|\mathsf{K}^{\psi}\|_F.$$

□

As a consequence, we have the following result concerning the SR.

Corollary 16.3.1 *The VSK κ_G^{Ψ} is more expressive than κ_G.*

Proof. First, we observe that for any set of nodes

$$\mathrm{tr}(\mathsf{K}^{\Psi}) = \mathrm{tr}(\mathsf{K}) = N\varphi_G(0) = N.$$

Then, taking into account Theorem 16.3.4, we obtain

$$SR(\mathsf{K}) = \frac{N}{\|\mathsf{K}\|_F} \leq \frac{N}{\|\mathsf{K}^{\Psi}\|_F} = SR(\mathsf{K}^{\Psi}).$$

□

We point out that, since the Gaussian VSK is more expressive than the standard one, the VSK setting may be able to handle more complex tasks.

16.3.2.2 Spectrum of the Gaussian VSK

The conditioning of a positive definite matrix can be expressed as the ratio of its largest and smallest eigenvalues, as discussed in Section 4.1. More precisely, in Section 4.2 we observed that the smallest eigenvalue is bounded from below in terms of the separation distance q_X (see Definition 4.1.2).

As reported in Proposition 9.1.1 in Section 9.1, we have that

$$q_X \leq q_X^{\Psi},$$

where

$$q_X^{\Psi} = \frac{1}{2} \min_{i \neq j} \|\Psi(\boldsymbol{x}_i) - \Psi(\boldsymbol{x}_j)\|,$$

is the separation distance in the VSK setting. For the Gaussian kernel, we can prove the following.

Proposition 16.3.2 *Let $X = \{\boldsymbol{x}_k,\ k = 1, \ldots, N\} \subset \Omega$ and let K^{Ψ} and K be as in Theorem 16.3.2. Then,*

$$\mathrm{cond}(\mathsf{K}^{\Psi}) \leq \mathrm{cond}(\mathsf{K}).$$

Proof. From Theorem 16.3.3, since $\mathsf{K}_{ii}^{\psi} = 1$, $i = 1, \ldots, N$, we obtain

$$\mathrm{cond}(\mathsf{K}^{\Psi}) = \frac{\lambda_{\max}(\mathsf{K}^{\Psi})}{\lambda_{\min}(\mathsf{K}^{\Psi})} \leq \frac{\lambda_{\max}(\mathsf{K})}{\lambda_{\min}(\mathsf{K}^{\Psi})} \leq \frac{\lambda_{\max}(\mathsf{K})}{\lambda_{\min}(\mathsf{K})} = \mathrm{cond}(\mathsf{K}).$$

□

16.3.3 A Scaling Function for KRNs

Let us recall the setting of Section 13.1, where we introduced KRNs. Let $X = \{\boldsymbol{x}_k,\ k = 1,\dots,N\} \subseteq \Omega$, where $\Omega \subseteq \mathbb{R}^d$ is a bounded set, be a set of distinct data, and let $y_k \in \mathbb{R}$, $i = 1,\dots,N$ be the associated measurements or outputs.

Here, concerning the choice of the scaling function $\psi : \Omega \longrightarrow \Lambda$, we can, for example, model the trend of the data restricting to a specific class of functions, i.e. a model $s : \Omega \times \mathbb{R}^l \longrightarrow \mathbb{R}$ depending on $\boldsymbol{x} \in \Omega$, and on ℓ parameters $\boldsymbol{\beta} = (\beta_1,\dots,\beta_\ell)$. To determine $\boldsymbol{\beta}$, we compute:

$$\boldsymbol{\beta}^* = \min_{\boldsymbol{\beta}\in\mathbb{R}^\ell} \sum_{i=1}^{N} (y_i - s(\boldsymbol{x}_i,\boldsymbol{\beta}))^2.$$

Then, we can exploit the so-constructed auxiliary regression model and define the function $\psi : \Omega \longrightarrow \Lambda$ as

$$\psi(\boldsymbol{x}) = s(\boldsymbol{x},\boldsymbol{\beta}^*).$$

Of course, this gives a recipe for choosing the scaling function, which is not unique. exactking of time series, one usually disposes of other available and correlated data sampled at the same locations; for instance, when studying the evolution of a tumor mass over time, other biomarkers are often available and can be taken into account (see, e.g., Stura et al., 2016). This available information can be incorporated into the scaling map by adding additional features; see Campi et al. (2021) for an application to COVID-19 data.

16.3.4 A Scaling Map for the Classification with SVMs

In this section, the variably scaled setting is considered in the context of ensemble learning (see Zhang and Ma, 2012, for an overview of this topic). More precisely, we construct the scaling function using another auxiliary classifier *mathlcalC*, thereby treating the variably scaled setting as a stacking technique.

In this regard, let us suppose that we are dealing with a classification task involving ℓ classes $\mathcal{C}_1,\dots,\mathcal{C}_\ell$, $\ell \in \mathbb{N}$. For the variably scaled scheme, we propose the map

$$\psi(\boldsymbol{x}) := (P_1(\boldsymbol{x}),\dots,P_{\ell-1}(\boldsymbol{x})), \tag{16.5}$$

where $P_k(\boldsymbol{x})$ is the estimated probability that the element $\boldsymbol{x} \in \Omega$ belongs to class $\mathcal{C}_k$ given by the auxiliary method.

We point out that

$$P_\ell(\boldsymbol{x}) = 1 - \sum_{k=1}^{\ell-1} P_k(\boldsymbol{x})$$

is not included in the map because it is correlated with the other features.

As far as the results in Section 16.3.1 are concerned, taking the linear kernel and the map proposed in (16.5) we obtain

$$\mathsf{K}_{ij}^{\psi} = \sum_{k=1}^{\ell-1} P_k(\boldsymbol{x}_i) P_k(\boldsymbol{x}_j).$$

Taking the Gaussian VSK (see Section 16.3.2), we have

$$\mathsf{K}_{ij}^{\psi} = \prod_{k=1}^{\ell-1} e^{-(P_k(\boldsymbol{x}_i) - P_k(\boldsymbol{x}_j))^2}.$$

Here, we observe that if $\psi(\boldsymbol{x}_i) \approx \psi(\boldsymbol{x}_j)$, then $\mathsf{K}_{ij}^{\psi} \approx 1$ and so $\mathsf{K}_{ij}^{\Psi} \approx \mathsf{K}_{ij}$.

The combination of SVMs with NB and other relevant classifiers such as kNN and random forests (see Section 13.3) found application in Marchetti et al. (2020), where the task is to predict the hormone treatment of human-derived xenografts from RNA sequencing data.

17

Kernels for Persistent Homology

Homology, and in particular persistent homology, is a tool used in topological data analysis (TDA) to get features of data, such as connected components, high-dimensional holes, and cavities. Simplicial complexes are used as a basis to construct the so-called persistence diagrams (PDs). We introduce kernels defined on PD and their extension, which we call variably scaled persistence kernels (VSPKs). Examples of their use in SVM classification of popular data sets in data analysis are discussed in this chapter.

17.1 Basics and Notations

Let $\Omega \subset \mathbb{R}^{\nu}$ and let $\mathscr{X} = \{\boldsymbol{x}_1, \ldots, \boldsymbol{x}_n\} \subset \Omega$ be a set of input data, $\nu, n \in \mathbb{N}$. In the classification setting, each element $\boldsymbol{x}_i$ of the dataset, $i = 1, \ldots, n$, is uniquely associated to a *label* (or *class*) y_i, which is an element of a certain set of labels $\mathscr{T} = \{c_1, \ldots, c_t\}$, $t \in \mathbb{N}$. In principle, the labels can be real numbers, but they can also represent categorical concepts to be learned. Then, we can define the dataset $\mathscr{L} = \{(\boldsymbol{x}_i, y_i) \,|\, \boldsymbol{x}_i \in \mathscr{X}, y_i \in \mathscr{T}\}$. The binary classification learning problem 12.1.1 of Section 12 can be generalized as follows.

Problem 17.1.1 *Find a decision function* $s : \Omega \longrightarrow \mathscr{T}$ *such that*

(i) it models the input-output relation in $\mathscr{L}$*;*

(ii) letting $\boldsymbol{\xi}_1, \ldots, \boldsymbol{\xi}_m \in \Omega \setminus \mathscr{X}$ *be unseen instances,* $m \in \mathbb{N}$*, with associated labels* $\iota_i \in \mathscr{T}$*, s models the input-output relation* $\{(\boldsymbol{\xi}_i, \iota_i)\}_{i=1,\ldots,m} \subset (\Omega \setminus \mathscr{X}) \times \mathscr{T}$*,* $m \geq 1$.

The *generalization* capability required in (ii) is fundamental in the learning problem, since it is trivial to find a decision function that satisfies (i); for an introduction concerning (statistical) learning theory, refer to, for example, Schölkopf and Smola (2002, §1.3) and Vapnik (1998).

Kernel methods are well-established tools in supervised machine learning and across various research and applied fields (Schaback and Wendland, 2006; Shawe-Taylor and Cristianini, 2004; Shawe-Taylor and Sun, 2014). The flexibility provided by kernel-based schemes allows for the handling of different kinds of structured data, for example, graphs and words, which are encoded in some dot product space where even complex patterns may be distinguished (cf., e.g., Borgwardt et al., 2005; Collins and Duffy, 2001; Lampert, 2009; Vishwanathan et al., 2010),

In this chapter, our data consist of *PDs*, which are outputs of the so-called *persistent homology* in the framework of TDA. The use of topological methods and analysis techniques to extract significant features and patterns from data is receiving increasing interest; for an overview of TDA, see, for example, Carlsson (2009); Edelsbrunner et al. (2000); and Fomenko (1994). We notice that in recent years, several kernels specifically designed to handle the peculiar structure of PDs have been proposed (Carrière et al., 2017; Kusano et al., 2017; Le and Yamada, 2018; Reininghaus et al., 2015); therefore, the construction of suitable kernels is a very active research line in this context.

The interactions between kernel methods, frame analysis, and persistent homology in the framework of RKHSs have been explored in Guillemard and Iske (2017). The authors showed that interactions among these novel methodologies create new opportunities for constructing numerical algorithms to analyze properties of data that have so far remained unexplored.

In this chapter, we introduce the primary tool that we named *VSPKs*, which are inspired, as we did in Chapter 9, by the VSK introduced in the context of approximation theory (see the fundamental reference Bozzini et al., 2015) and employed as a feature augmentation strategy in Campi et al. (2021) and Marchetti et al. (2020) for kernel-based learning. Indeed, we aim to build a bridge between classical VSKs and kernels defined in the context of persistent homology. After defining VSPKs, we design scaling functions and test the performance of the resulting kernels across three datasets. The results show that the variably scaled setting may yield better classification performance and efficiency than the standard setting, and thus warrants further investigation and application in persistence diagram-based settings.

17.2 Positive Definite and Variably Scaled Kernels

Let $\kappa : \Omega \times \Omega \longrightarrow \mathbb{R}$ be a kernel. Given a set of data $\mathscr{X} = \{\boldsymbol{x}_1, \ldots, \boldsymbol{x}_n\} \subset \Omega$, the $n \times n$ matrix K with elements $K_{ij} = \kappa(\boldsymbol{x}_i, \boldsymbol{x}_j)$, $i, j = 1, \ldots, n$, is the *Gram matrix* of the kernel κ with respect to $\mathscr{X}$. If κ is positive definite (strictly positive definite) on $\Omega \times \Omega$, that is, K is positive semi-definite (definite) for all possible datasets in Ω. It is possible to decompose the kernel according to *Mercer's Theorem* (Mercer, 1909) and to interpret such decomposition as an inner product in a RKHS $\mathscr{F}$. Indeed, there exists a (nonunique) *feature map* $\Phi : \Omega \longrightarrow \mathscr{F}$ such that

$$\kappa(\boldsymbol{x}, \boldsymbol{y}) = \langle \Phi(\boldsymbol{x}), \Phi(\boldsymbol{y}) \rangle_{\mathscr{F}} \qquad \boldsymbol{x}, \boldsymbol{y} \in \Omega,$$

being $\langle \cdot, \cdot \rangle_{\mathscr{F}}$ the bilinear form related to the RKHS (e.g. $\Phi(\boldsymbol{x}) = \kappa(\cdot, \boldsymbol{x})$). Moreover, the kernel κ induces a distance d_κ on Ω

$$d_\kappa(\boldsymbol{x}, \boldsymbol{y}) = \kappa(\boldsymbol{x}, \boldsymbol{x}) + \kappa(\boldsymbol{y}, \boldsymbol{y}) - 2\kappa(\boldsymbol{x}, \boldsymbol{y}). \qquad (17.1)$$

As already mentioned, VSKs were introduced in Bozzini et al. (2015) in the context of kernel-based approximation to overcome instability issues. Then, they have been extended to work in a more general setting in Campi et al. (2021).

Definition 17.2.1 Let $\Lambda \subseteq \mathbb{R}^\nu$, $\nu > 0 \in \mathbb{N}$ and let $\kappa : \tilde{\Omega} \times \tilde{\Omega} \longrightarrow \mathbb{R}$ be a continuous (strictly) positive definite kernel, where $\tilde{\Omega} = \Omega \times \Lambda \subseteq \mathbb{R}^{\nu+\nu}$. Given a *scaling* function $\psi : \Omega \longrightarrow \Lambda$, and by defining $\Psi(\boldsymbol{x}) = (\boldsymbol{x}, \psi(\boldsymbol{x}))$, a VSK $\kappa^\Psi : \Omega \times \Omega \longrightarrow \mathbb{R}$ is defined as

$$\kappa^\Psi(\boldsymbol{x}, \boldsymbol{y}) = \kappa((\boldsymbol{x}, \psi(\boldsymbol{x})), (\boldsymbol{y}, \psi(\boldsymbol{y})))$$

for $\boldsymbol{x}, \boldsymbol{y} \in \Omega$.

The function Ψ can be interpreted as a *feature augmentation* map, which adds ν coordinates (features) to the original sample. In this view, the VSK setting has been analyzed in Campi et al. (2021) as a *stacking* technique that enhances the predictive performance of classical kernel-based classifiers, such as SVMs.

Letting $\boldsymbol{x} = (x_1, \ldots, x_d)^\intercal \in \Omega$, we recall that a binary (i.e., $\mathscr{T} = \{-1, +1\}$) SVMs classifier is characterized by the decision function

$$s(\boldsymbol{x}) = \text{sign}(h(\boldsymbol{x})) = \text{sign}(\langle \Phi(\boldsymbol{x}), \boldsymbol{w} \rangle_{\mathscr{F}} + b),$$

where

$$\boldsymbol{w} = \sum_{i=1}^{n} \alpha_i y_i \Phi(\boldsymbol{x}_i) \in \mathscr{F}.$$

The coefficients $\boldsymbol{\alpha} = (\alpha_1, \ldots, \alpha_n) \in \mathbb{R}^n$ are the solution of the following *soft margin* problem (Fasshauer and McCourt, 2015, §18, pp. 346–347)

$$\begin{cases} \min_{\boldsymbol{\alpha} \in \mathbb{R}^n} \frac{1}{2} \sum_{i=1}^n \sum_{j=1}^n \alpha_i \alpha_j y_i y_j \kappa(\boldsymbol{x}_i, \boldsymbol{x}_j) - \sum_{i=1}^n \alpha_i, \\ \text{s.t. } \sum_{i=1}^n \alpha_i y_i = 0, \\ 0 \le \alpha_i \le \zeta, \quad i = 1, \ldots, n, \end{cases}$$

where $[0, \zeta]^n$ is the *bounding box*, with $\zeta \in [0, +\infty)$. Usually, a binary SVM classifier is extended to the multiclass setting by considering a *one-vs-rest* approach. We recall that details on hard and soft margin problems have been discussed in Chapter 13.

The strength of kernel-based classifiers, such as SVMs, largely depends on the flexibility of kernel construction. Indeed, any structured data may potentially be *translated* into a particular Hilbert space (e.g., the classical $\mathbb{R}^n$) by an appropriately defined feature map. This is the case of *PDs*, which are valuable tools in TDA characterized by a peculiar structure, as we outline in Section 17.3.

17.3 Persistent Homology and Kernels

We start the session by introducing some concepts about homology. This procedure associates a sequence of abelian groups (or modules) to a given mathematical object, as well as *persistent homology*. For more details, especially concerning the algebraic aspects of the construction, we invite the readers to look at Edelsbrunner et al. (2000); Guillemard and Iske (2017); and De Marchi (n.d.).

17.3.1 Basics on Persistent Homology

Let our data $\mathscr{X} = \{\boldsymbol{x}_1, \ldots, \boldsymbol{x}_n\} \subset \Omega$ be interpreted as a set of vertices sampled from some manifold M, and suppose that we wish to highlight some of its intrinsic homological properties. Intuitively, such properties provide important geometrical information on $\mathscr{X}$ by describing the number of connected components, holes, and cavities present in the data distribution. A way to study these geometrical features is to consider the manifold built on $\mathscr{X}$, which we treat as an approximation of M. More precisely, given $\varepsilon > 0$, we study

$$M^{\mathscr{X}, \varepsilon} = \bigcup_{i=1}^{n} B(\boldsymbol{x}_i, \varepsilon),$$

where $B(\boldsymbol{x}_i, \varepsilon)$ is the ball centered in $\boldsymbol{x}_i$ of radius ε. Then, we associate to $M^{\mathscr{X},\varepsilon}$ an abstract *simplicial complex*, which is a family of sets closed with respect to the *subset operator*. In a sense, it can be considered as an enriched graph structure, where not only points and edges are included but also triangles (i.e., formal combinations of three points in two dimensions), tetrahedra (i.e., formal combinations of four points in three dimensions), and so on (a generalization of triangulations of a topological space) (see Figure 17.1). Moreover, the concept of *face* plays a central role. For example, an edge is characterized by two zero-dimensional faces (two extremal points), a triangle by three 1D faces (three-side edges), and so on. A simple definition is Definition 17.3.1.

Definition 17.3.1 A simplicial complex K consists of a set of simplices of certain dimensions that have to meet the following conditions:

(i) Every face of a simplex in K is also in K.
(ii) The non-empty intersection of any two simplices $\sigma_1, \sigma_2 \in K$ is a face of both the simplices.
(iii) The dimension of the simplicial complex K is the maximum dimension of simplices that belong to it.

Simplicial complexes can be defined by different construction rules that determine how higher-dimensional structures are built from lower-dimensional ones. An example of a simplicial complex is provided in Figure 17.1 Here, we associate to $M^{\mathscr{X},\varepsilon}$ a *Vietoris-Rips simplicial complex* $K^{\mathscr{X},\varepsilon}$, where two distinct vertices $\boldsymbol{x}_i, \boldsymbol{x}_j$ are connected by an edge if and only if $\|\boldsymbol{x}_i - \boldsymbol{x}_j\|_2 \leq \varepsilon$, and we consider r-dimensional elements that are determined by $r+1$ connected $(r-1)$-dimensional faces as long as $r \leq v$. In Figure 17.4a, 17.4b, and 17.4c, we show a 2D example ($v = 2$) where different numbers of edges and triangles are generated depending on the chosen ε.

On the Abelian group $\mathbb{G} = (\mathbb{Z}, +)$ we can consider linear combinations of simplices with coefficients in $\mathbb{G}$, getting *chains of simplices*. Integer combinations of r-dimensional simplices in $K^{\mathscr{X},\varepsilon}$ form the *r-chains* algebraic group $C_r^{\mathscr{X},\varepsilon}$.

- σ and $-\sigma$ are the simpliest chains.
- Let $\sigma_i, i = 1,2,3$ be 1-simplices (i.e., line segments), then $2\sigma_1 - \sigma_2 + 2\sigma_3$ is a 1-chain, that is an element of $C_1^{\mathscr{X},\varepsilon}$. The coefficients indicate how many times each simplex is included with its orientation (positive/negative).

Moreover, letting $[\boldsymbol{x}_{i_0}, \ldots, \boldsymbol{x}_{i_r}]$ be the face constructed upon distinct vertices

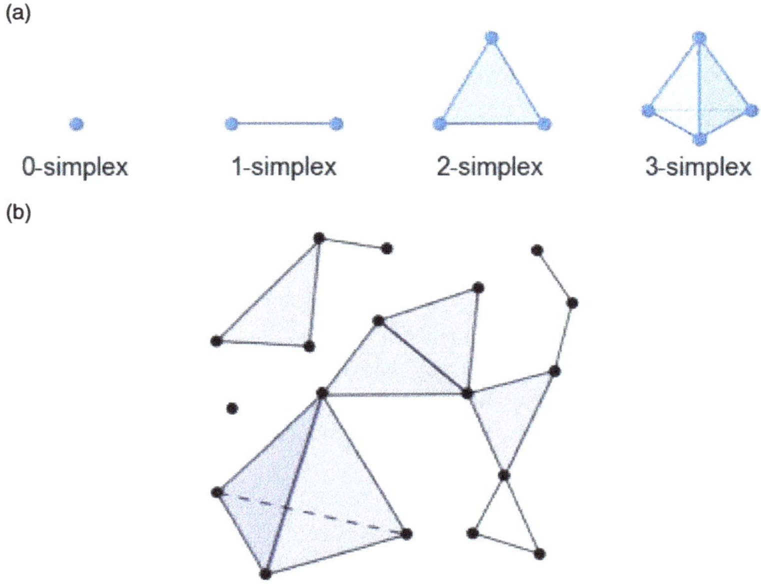

Figure 17.1 (a) Simplices up to dimension 3. (b) A simplicial complex of dimension 3.

$\boldsymbol{x}_{i_0},\ldots,\boldsymbol{x}_{i_r}$ in $\mathscr{X}$, $i_0,\ldots,i_r \in \{1,\ldots,n\}$ (which is an r-dimensional simplex), we consider the *Boundary operator* $\partial_r : C_r^{\mathscr{X},\varepsilon} \longrightarrow C_{r-1}^{\mathscr{X},\varepsilon}$ as

$$\partial_r[\boldsymbol{x}_{i_0},\ldots,\boldsymbol{x}_{i_r}] = \sum_{j=0}^{r}(-1)^r[\boldsymbol{x}_{i_0},\ldots,\boldsymbol{x}_{i_{j-1}},\boldsymbol{x}_{i_{j+1}},\ldots,\boldsymbol{x}_{i_r}].$$

In practice, the boundary of a chain is the linear combination of the boundary of the simplices in the r-chain, which is a chain of order $r-1$. The operator ∂_r is linear and $\partial_r^2 = 0$, since the boundary of a simplex has no boundary. Moreover, if $\partial_r(ac) = 0$ with $a \neq 0$, then $\partial_r = 0$.

Example 17.3.1 *Consider the path from the point $\boldsymbol{x}_1$ to $\boldsymbol{x}_4$. Take the 1-simplices $\sigma_1 = [\boldsymbol{x}_1,\boldsymbol{x}_2]$, $\sigma_2 = [\boldsymbol{x}_2,\boldsymbol{x}_3]$, $\sigma_3 = [\boldsymbol{x}_3,\boldsymbol{x}_4]$, and the chain $c = \sigma_1+\sigma_2+\sigma_3$. Hence,*

$$\begin{aligned}\partial_1 c &= \partial_1(\sigma_1+\sigma_2+\sigma_3) = \partial_1(\sigma_1)+\partial_1(\sigma_2)+\partial_1(\sigma_3),\\ &= \partial_1([\boldsymbol{x}_1,\boldsymbol{x}_2])+\partial_1([\boldsymbol{x}_2,\boldsymbol{x}_3])+\partial_1([\boldsymbol{x}_3,\boldsymbol{x}_4])\\ &= (\boldsymbol{x}_2-\boldsymbol{x}_1)+(\boldsymbol{x}_3-\boldsymbol{x}_2)(\boldsymbol{x}_4-\boldsymbol{x}_3) = \boldsymbol{x}_4-\boldsymbol{x}_3.\end{aligned}$$

Let c an r-chain. It is called an *r-cycle* if its boundary is zero, that is, $\partial c = 0$. As an example, consider a closed polygonal curve. Boundaries are cycles (not the opposite!), so chains form a chain complex, whose homology groups (cycles modulo boundaries) are called *simplicial homology groups*.

More specifically, *r-cycles* and *r-boundaries* form Abelian groups that are denoted as $Z_r^{\mathscr{X},\varepsilon} = \ker\partial_r$ and $B_r^{\mathscr{X},\varepsilon} = \mathrm{im}\partial_{r+1}$, respectively. Furthermore, it can be proved that the rank of the *r-homology* group $H_r^{\mathscr{X},\varepsilon} = Z_r^{\mathscr{X},\varepsilon}/B_r^{\mathscr{X},\varepsilon}$ encapsulates the concept of *r-dimensional holes* in set of the complexes $K^{\mathscr{X},\varepsilon}$, thus relating M and $\mathscr{X}$, for characterizing their geometrical features.

Unfortunately, finding an optimal $\varepsilon^\star$ that represents the intrinsic geometric properties of M, is a tough and unstable process. This is evident in the example depicted in Figure 17.4a, 17.4b, and 17.4c, where different values of ε determine a completely different simplicial structure.

To solve this issue, one can drop the idea of selecting a good parameter and analyze the whole *filtration* $\{M^{\mathscr{X},\varepsilon} \mid \varepsilon > 0\}$.

Definition 17.3.2 Given the simplicial complex $\mathscr{K}$,a family $(K_t)_{t\in T}$ of sub-complexes, where T is a totally ordered set is a **filtration** if:

- $\forall t_1, t_2 \in T$ with $t_1 < t_2$ we get $K_{t_1} \subset K_{t_2}$
- $\bigcup_{t\in T} K_t = \mathscr{K}$

The simplicial complex $\mathscr{K}$ obtained with such a sequence of subcomplexes is termed *filtered simplicial complex*. In Figure 17.2 we show two filtrations of the complex consisting of the union of two touching triangles (figures taken from Horak et al., 2009, Figs. 9 and 10), while in Figure 17.3 the evolving sequence of complexes (figure taken from Otter et al., 2017, Fig. 5).

In particular, letting $\varepsilon_1 < \cdots < \varepsilon_u$ be increasing real numbers, the nested sequence $K_{\mathscr{X},\varepsilon_1} \subseteq \cdots \subseteq K_{\mathscr{X},\varepsilon_u}$ is obtained. Then, for $r \geq 0$ and $i \in \{1,\ldots,u\}$, we can associate to this filtration the *ℓ-persistent homology* group

$$H_{r,\ell}^{\mathscr{X},\varepsilon_i} = Z_r^{\mathscr{X},\varepsilon_i}/(Z_r^{\mathscr{X},\varepsilon_i} \cap B_r^{\mathscr{X},\varepsilon_{i+\ell}}).$$

The group $H_{r,\ell}^{\mathscr{X},\varepsilon_i}$ contains the homology classes that persist in the *time interval* $[i, i+\ell]$ that is, they are *born* at $K_{\mathscr{X},\varepsilon_j}$ for some $j < i$ and *die* at $K_{\mathscr{X},\varepsilon_{i+\ell}}$. Such homology classes might persist *indefinitely* (this is denoted as ∞) or might *die* with a certain ε_j, $i < j \leq u$. Therefore, studying persistent homology groups provides information about the connected components, holes, and cavities in the data, along with their persistence. The more persistent an element is, the more likely it is to reflect the data structure.

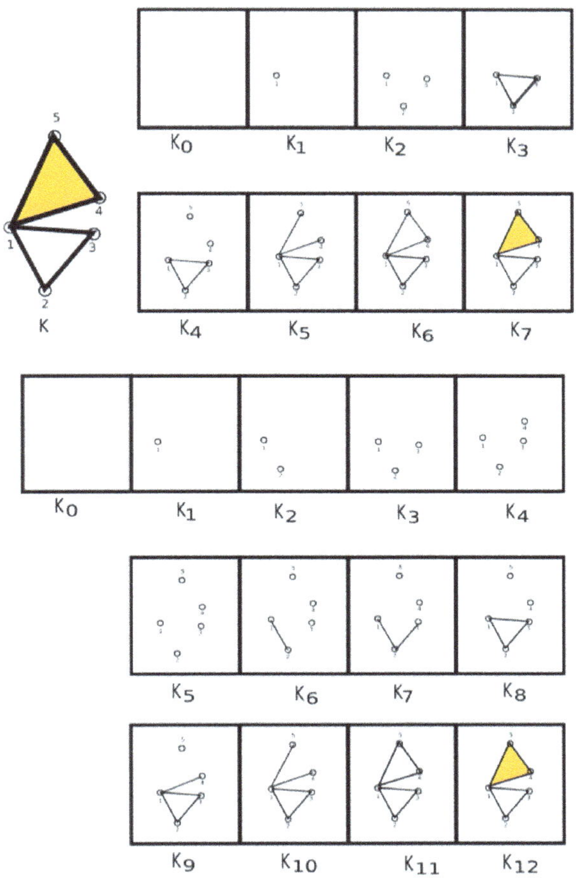

Figure 17.2 Two filtrations of the complex made up of two touching triangles.

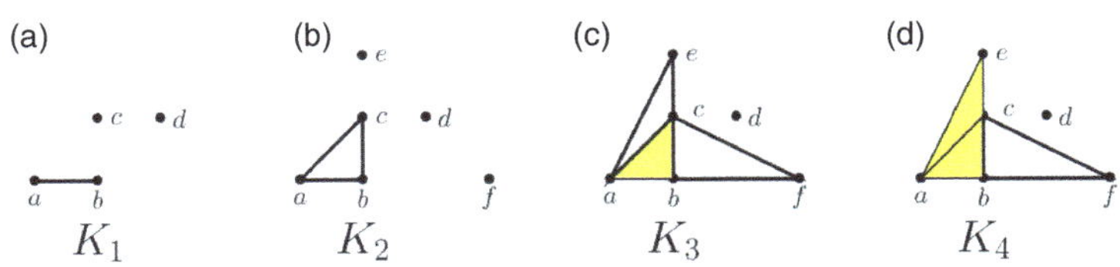

Figure 17.3 (a–d) Example of a filtered complex.

Each element of the persistent homology groups obtained by considering the whole filtration can be represented by a birth-death pair $(b,d) \in \mathbb{R}^2_+, b = \varepsilon_h, d = \varepsilon_k$ for some $h \in \{1,\dots,m\}, k \in \{1,\dots,m\} \cup \{\infty\}, h < k$. We say that $d-b$ is the *persistence* of (b,d), and that a birth-death pair is r-dimensional if it is related to r-dimensional homology groups.

Definition 17.3.3 Letting $\boldsymbol{\varepsilon} = (\varepsilon_1,\dots,\varepsilon_u)$ a *persistence diagram* $D_r(\mathscr{X},\boldsymbol{\varepsilon})$ related to the filtration $K^{\mathscr{X},\varepsilon_1} \subseteq \cdots \subseteq K^{\mathscr{X},\varepsilon_m}$ is the multiset

$$D_r(\mathscr{X},\boldsymbol{\varepsilon}) = \{(b,d) \mid (b,d) \in P_r(\mathscr{X},\boldsymbol{\varepsilon})\} \cup \mathscr{B}, \quad r \geq 0, \tag{17.2}$$

subset of $\mathbb{R}^2$, where $P_r(\mathscr{X},\boldsymbol{\varepsilon})$ denotes the set of r-dimensional birth-death pairs obtained with the filtration and $\mathscr{B} = \{(x,x) \mid x \geq 0\}$.

We remark that $D_r(\mathscr{X},\boldsymbol{\varepsilon})$ is a multiset, since a couple (b,d) might appear more than once, that is, might have multiplicity greater than one. Furthermore, the bisector $\mathscr{B} = \{(x,x), x \geq 0\}$ is composed of an infinite number of elements of infinite multiplicity, which are concerned with noise in the data. It is added to achieve *uniformity* across different PDs and to facilitate the formulation of proper metrics, as we present below. Each point in $(b,d) \in D_r(\mathscr{X},\boldsymbol{\varepsilon})$ is called *generator* and the difference $d-b$ its *persistence*.

In Figure 17.4d, we display the result of the discussed analysis with the example dataset (17.4a). Such analysis captures some intrinsic geometrical properties of $\mathscr{X}$. In particular, the unique most persistent zero-dimensional pair, which is significantly far from the others, suggests that a unique connected component underlies $\mathscr{X}$. Moreover, the persistent 1D pair indicates a one-dimensional hole in the dataset's structure, as highlighted by the simplicial complex shown in the 17.4c panel.

Persistence diagrams exhibit stability under perturbations of the underlying dataset, as discussed in Cohen-Steiner et al. (2007). To clarify this point, we recall two useful metrics for comparing datasets and PDs.

Definition 17.3.4 Let $\mathscr{X},\mathscr{Y} \subset \Omega$ be two non-empty datasets. The *Hausdorff distance* between $\mathscr{X},\mathscr{Y}$ is

$$d_H(\mathscr{X},\mathscr{Y}) = \max\left\{ \sup_{x\in\mathscr{X}} \inf_{y\in\mathscr{Y}} \|x-y\|_\infty, \sup_{y\in\mathscr{Y}} \inf_{x\in\mathscr{X}} \|x-y\|_\infty \right\}.$$

Definition 17.3.5 Let $D_r(\mathscr{X},\boldsymbol{\varepsilon})$ and $D_r(\mathscr{Y},\boldsymbol{\varepsilon})$ be PDs for some $r \geq 0$ and filtration radii vector $\boldsymbol{\varepsilon}$. The *p-Wasserstein distance*

$$d_{W,p}(D_r(\mathscr{X},\boldsymbol{\varepsilon}),D_r(\mathscr{Y},\boldsymbol{\varepsilon})) = \left(\inf_{\gamma\in\Gamma} \sum_{x\in D_r(\mathscr{X},\boldsymbol{\varepsilon})} \|x-\gamma(x)\|_\infty^p \right)^{\frac{1}{p}},$$

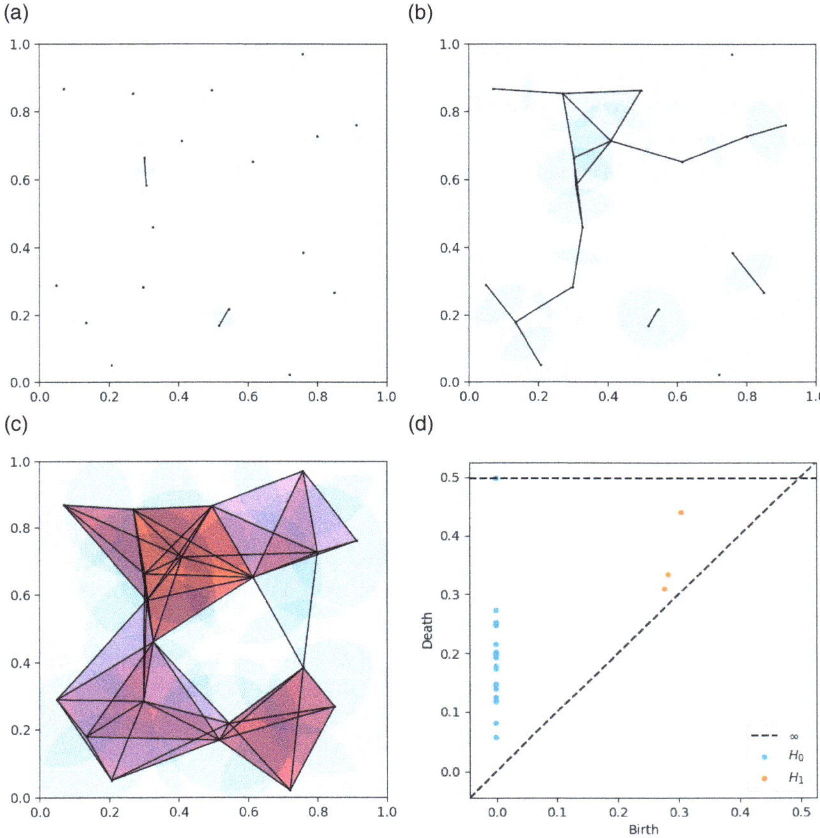

Figure 17.4 Given a set of points $\mathscr{X}$, the construction of the Vietoris–Rips complex along the filtration is depicted in (a) $\varepsilon = 0.1$; (b) $\varepsilon = 0.22$ and (c) $\varepsilon = 0.4$. The persistence diagrams $D_0(\mathscr{X}, \boldsymbol{\varepsilon})$ and $D_1(\mathscr{X}, \boldsymbol{\varepsilon})$ are overlapped in (d).

where $\Gamma = \{\gamma : D_r(\mathscr{X}, \boldsymbol{\varepsilon}) \longrightarrow D_r(\mathscr{Y}, \boldsymbol{\varepsilon}) \mid \gamma \text{ is a bijection}\}$. In particular, letting $p \to \infty$, we obtain the *bottleneck distance*

$$\begin{aligned} d_{W,\infty}(D_r(\mathscr{X}, \boldsymbol{\varepsilon}), D_r(\mathscr{Y}, \boldsymbol{\varepsilon})) &= d_B(D_r(\mathscr{X}, \boldsymbol{\varepsilon}), D_r(\mathscr{Y}, \boldsymbol{\varepsilon})) \\ &= \inf_{\gamma \in \Gamma} \sup_{x \in D_r(\mathscr{X}, \boldsymbol{\varepsilon})} \|x - \gamma(x)\|_\infty . \end{aligned}$$

We have the stability result Kusano et al. (2017)

$$d_B(D_r(\mathscr{X}, \boldsymbol{\varepsilon}), D_r(\mathscr{Y}, \boldsymbol{\varepsilon})) \leq d_H(\mathscr{X}, \mathscr{Y}),$$

that is, the bottleneck distance between PDs is controlled as long as the underlying datasets are close in the Hausdorff metric.

For readers interested in computing these distances using Python, the module `gudhi` contains the function `gudhi.bottleneck_distance` that works on PDs. The Wasserstein distance between persistence diagrams can be computed using the function `wasserstein_distance_pd` available in the `scipy.stats` module.

17.3.2 Kernels for Persistence Diagrams

To better measure similarities between PDs, various positive-definite kernels suitable for their peculiar structure have been introduced and studied in the recent literature. In our applications, we will consider the following ones. Besides, we will sometimes use the notations $D_1 = D_r(\mathscr{X}, \boldsymbol{\varepsilon})$ and $D_2 = D_r(\mathscr{Y}, \boldsymbol{\varepsilon})$ for simplicity.

- The **Persistence Scale Space (PSS) kernel** Reininghaus et al. (2015).

  The key idea is to compute the feature map as the solution of the *heat equation*. We consider $\Omega_{ad} = \{\mathbf{x} = (x_1, x_2) \in \mathbb{R}^2 : x_2 \geqslant x_1\}$ and denote with $\delta_{\mathbf{x}}$ the Dirac delta centered at $\mathbf{x}$. For a given $D \in \mathfrak{D}_r$, we consider the solution $u : \Omega_{ad} \times \mathbb{R}_{\geqslant 0} \to \mathbb{R}$, $(\mathbf{x}, t) \mapsto u(\mathbf{x}, t)$ of the following PDE:

$$\begin{aligned} \Delta_{\mathbf{x}} u &= \partial_t u \quad \text{in } \Omega_{ad} \times \mathbb{R}_{\geqslant 0} \\ u &= 0 \quad \text{on } \partial\Omega_{ad} \times \mathbb{R}_{\geqslant 0} \\ u &= \sum_{\mathbf{y} \in D} \delta_{\mathbf{y}} \quad \text{on } \Omega_{ad} \times 0. \end{aligned}$$

  The feature map $\Phi_\sigma : \mathfrak{D}_r \to \mathrm{L}^2(\Omega_{ad})$ at scale $\sigma > 0$ at the persistence diagram D is defined as $\Phi_\sigma(D) = u\big|_{t=\sigma}$. This map yields the **Persistence Scale-Space Kernel (PSS kernel)** K_{PSS} on $\mathfrak{D}_r$ as:

$$\kappa_\sigma(D_1, D_2) = \langle \Phi_\sigma(D_1), \Phi_\sigma(D_2) \rangle_{\mathrm{L}^2(\Omega_{ad})}.$$

  We may write it explicitly as follows.

$$\kappa_\sigma(D_1, D_2) = \frac{1}{8\pi\sigma} \sum_{\substack{y \in D_1 \\ z \in D_2}} \exp\left(-\frac{||\mathbf{y} - \mathbf{z}||^2}{8\sigma}\right) - \exp\left(-\frac{||\mathbf{y} - \overline{\mathbf{z}}||^2}{8\sigma}\right), \tag{17.3}$$

  where $\overline{\mathbf{z}} = (d, b)$ if $\mathbf{z} = (b, d)$.

  The kernel κ_σ is 1-Wasserstein stable, i.e.:

$$||\Phi_\sigma(D_1) - \Phi_\sigma(D_2)||_{L_2(\Omega)} \leq \frac{1}{2\sqrt{\pi}\sigma} d_{W,1}(D_1, D_2),$$

- The **Persistence Weighted Gaussian kernel (PWG)** Kusano et al. (2017)

  Here, the idea is to replace each persistence diagram with a discrete measure. Starting with a strictly positive definite kernel, as the Gaussian one $\kappa_G(\mathbf{x},\mathbf{y}) = e^{-\frac{\|\mathbf{x}-\mathbf{y}\|^2}{2\sigma^2}}$, $\sigma > 0$ we denote the corresponding RKHS $\mathscr{H}_{\kappa_G}$.

  If $\Omega \subset \mathbb{R}^d$, we denote with $M_b(\Omega)$ the space of finite signed Radon measures and

$$E_{\kappa_G} : M_b(\Omega) \to \mathscr{H}_{\kappa_G}, \mu \mapsto \int_\Omega \kappa_G(\cdot,\mathbf{x})d\mu(\mathbf{x}).$$

Then, the PWG kernel is defined as

$$\kappa_G(D_1,D_2;\kappa,\omega) = \exp\left(-\frac{1}{2\tau^2}\|E_\kappa(\mu^\omega_{D_1}) - E_\kappa(\mu^\omega_{D_2})\|^2_{\mathscr{F}}\right) \quad \tau > 0, \tag{17.4}$$

which is built upon a *standard* Gaussian kernel κ and a weight function ω, where

$$E_\kappa(\mu^\omega_{D_1}) = \sum_{\boldsymbol{x}\in D_1} \omega(\boldsymbol{x})\,\kappa(\cdot,\boldsymbol{x}).$$

It is both 1-Wasserstein and bottleneck stable if we choose as a weight function

$$\omega_{\text{arc}}(\boldsymbol{x}) = \arctan(C(d-b)^\delta) \quad \boldsymbol{x} = (b,d), C > 0, \delta \in \mathbb{Z}_{>0},$$

where $d-b$ is the persistence. Indeed, there exist $\delta \in \mathbb{Z}_{>0}$ and $L > 0$ such that

$$\|E_\kappa(\mu^{\omega_{\text{arc}}}_{D_1}) - E_\kappa(\mu^{\omega_{\text{arc}}}_{D_2})\| \le L d_B(D_1,D_2). \tag{17.5}$$

- The **Sliced Wasserstein (SW) kernel** Carrière et al. (2017)

  We consider μ and ν two nonnegative measures, not necessarily normalized, on $\mathbb{R}$ such that $\mu(\mathbb{R}) = r = |\mu|$ and $\nu(\mathbb{R}) = r = |\nu|$, we recall that the 1-Wasserstein distance for nonnegative measures

$$\mathscr{W}(\mu,\nu) := \inf_{P\in\Pi(\mu,\nu)} \int\int_{\mathbb{R}\times\mathbb{R}} |x-y|dP(x,y)$$

where $\Pi(\mu,\nu)$ is the set of measures on $\mathbb{R}^2$ with marginals μ and ν.

Definition 17.3.6 Given $\theta \in \mathbb{R}^2$ with $\|\theta\|_2 = 1$, let $L(\theta)$ denote the line $\{\lambda\theta|\lambda \in \mathbb{R}\}$ and let $\pi_\theta : \mathbb{R}^2 \to L(\theta)$ be the orthogonal projection onto $L(\theta)$. Let $D_1, D_2 \in \mathfrak{D}_r$, and let $\mu^\theta_{D_1} := \sum_{\mathbf{x}\in D} \delta_{\pi_\theta(\mathbf{x})}$ and $\mu^\theta_{D_1\Delta} := \sum_{\mathbf{x}\in D} \delta_{\pi_\theta\circ\pi_\Delta(\mathbf{x})}$ and similarly for $\mu^{\theta}_{D_2}$ and $\mu^{\theta}_{D_2\Delta}$ where π_Δ is the orthogonal projection onto the diagonal. Then, the **Sliced Wasserstein (SW) distance** is

$$SW(D_1,D_2) = \frac{1}{2\pi}\int_{\mathbb{S}_1} \mathscr{W}(\mu^{\theta}_{D_1} + \mu^{\theta}_{D_2\Delta}, \mu^{\theta}_{D_2} + \mu^{\theta}_{D_1\Delta})d\theta.$$

with S_1 the usual $1d$ sphere, that is, the cicle.

Hence, the SW kernel is defined as follows

$$\kappa_{SW}(D_1,D_2) := \exp\left(-\frac{SW(D_1,D_2)}{2\sigma^2}\right). \tag{17.6}$$

Moreover, the SW distance, $SW(\cdot,\cdot)$, is equivalent to the 1-Wasserstein distance thanks to the following inequalities

$$\frac{1}{2M}d_{W,1}(D_1,D_2) \leq d_{SW}(D_1,D_2) \leq 2\sqrt{2}d_{W,1}(D_1,D_2),$$

for some positive constant M.

Remark 17.3.1 *We notice that the PWG and SW kernels are in the form*

$$\kappa(D_1,D_2) = \exp(-\beta d(D_1,D_2))$$

for some $\beta > 0$, where $d(\cdot,\cdot)$ is the distance induced by the underlying metric in the case of the SW, while for the PWG kernel the distance is induced by the kernel (see the definition of d_κ in (17.1)*).*

Other persistence kernels have been studied in the recent literature. It is worth mentioning the *Persistence Fisher Kernel*, which is a Riemannian manifold kernel for PDs (Le and Yamada, 2018), and the *Persistence Image Kernel*, for homological image analysis (Adams et al., 2017). These two kernels have been used, for instance, in SVM classification applications in (Bandiziol and De Marchi, 2024).

17.4 Variably Scaled Persistence Kernels

The purpose is to reinterpret the idea underlying VSKs (Bozzini et al., 2015) in the context of persistent homology. The main difference between standard kernels and persistence diagram kernels is the structure of the input data. Since PDs consists of collections of topological features, birth-death couples in $\mathbb{R}^2_{\geq\mathscr{B}} = \{\boldsymbol{x} = (b,d) \in \mathbb{R}^2_+ \mid d \geq b\}$, introducing a scaling function whose output lies outside $\mathbb{R}^2_{\geq\mathscr{B}}$ would be meaningless. Hence, letting $\mathfrak{D}_r(\boldsymbol{\varepsilon}) = \{D_r(\mathscr{X},\boldsymbol{\varepsilon}) \mid \mathscr{X} \subset \Omega\}$, we can make Definition 17.4.1.

Definition 17.4.1 Let $\kappa : \mathfrak{D}_r(\boldsymbol{\varepsilon}) \times \mathfrak{D}_r(\boldsymbol{\varepsilon}) \longrightarrow \mathbb{R}$ be a kernel for PDs and let $\Psi : \mathfrak{D}_r(\boldsymbol{\varepsilon}) \to \mathfrak{D}_r(\boldsymbol{\varepsilon})$. A variably scaled persistence kernel κ^{Ψ} on $\mathfrak{D}_r(\boldsymbol{\varepsilon}) \times \mathfrak{D}_r(\boldsymbol{\varepsilon})$ is

$$\kappa^{\Psi}(D_r(\mathscr{X},\boldsymbol{\varepsilon}), D_r(\mathscr{Y},\boldsymbol{\varepsilon})) := \kappa(\Psi(D_r(\mathscr{X},\boldsymbol{\varepsilon})), \Psi(D_r(\mathscr{Y},\boldsymbol{\varepsilon})))$$

for any couple of PDs $D_r(\mathscr{X},\boldsymbol{\varepsilon}), D_r(\mathscr{Y},\boldsymbol{\varepsilon}) \in \mathfrak{D}_r(\boldsymbol{\varepsilon})$.

As previously discussed, a proper function Ψ needs to be designed. We will construct it upon an auxiliary function $\psi : \mathfrak{D}_r(\boldsymbol{\varepsilon}) \longrightarrow \mathbb{R}^2_{\geq \mathscr{B}}$, whose design is deepened below in Section 17.4.3. Instead, we present two possible functions Ψ, that we denote as Ψ_a and Ψ_ρ.

17.4.1 The Feature Augmentation Map: Ψ_a

We define

$$\Psi_a(D_r(\mathscr{X},\boldsymbol{\varepsilon})) = D_r(\mathscr{X},\boldsymbol{\varepsilon}) \cup \psi(D_r(\mathscr{X},\boldsymbol{\varepsilon})).$$

The map Ψ_a plays the role of a *feature augmenting* map, since an additional generator is added in the persistence diagram. The following stability results holds for VSPKs with the map Ψ_a in terms of the classical setting (cf. De Marchi et al., 2022)

Proposition 17.4.1 *Let κ_σ be the PSS kernel defined in* (17.3) *and let D_1, D_2 be r-dimensional PDs for some $r \geq 0$. We have*

$$\|\Phi_\sigma^{\Psi_a}(D_1) - \Phi_\sigma^{\Psi_a}(D_2)\|_{L_2(\Omega)} \leq \frac{1}{2\sqrt{\pi}\sigma}\left(d_{W,1}(D_1,D_2) + \|\psi(D_1) - \psi(D_2)\|_\infty\right).$$

Proof. By definition, for any $\delta > d_{W,1}(D_1,D_2)$ there exists a bijection γ between D_1 and D_2 such that $\sum_{\boldsymbol{x}\in D_1} \|\boldsymbol{x} - \gamma(\boldsymbol{x})\|_\infty \leq \delta$. Then, by denoting $u_{\boldsymbol{y}}(\boldsymbol{x}) = \frac{1}{4\pi\sigma}\exp\left(-\frac{\|\boldsymbol{x}-\boldsymbol{y}\|^2}{4\sigma}\right)$ and imposing the extension $\gamma(\psi(D_1)) = \psi(D_2)$, we obtain

$$\begin{aligned}
\|\Phi_\sigma^{\Psi_a}(D_1) - \Phi_\sigma^{\Psi_a}(D_2)\|_{L_2(\Omega)} &= \|\Phi_\sigma(D_1 \cup \psi(D_1)) - \Phi_\sigma(D_2 \cup \psi(D_2))\| \\
&= \left\| \sum_{\boldsymbol{y}\in D_1 \cup \psi(D_1)} (u_{\boldsymbol{y}} - u_{\overline{\boldsymbol{y}}}) - (u_{\gamma(\boldsymbol{y})} - u_{\overline{\gamma(\boldsymbol{y})}}) \right\|_{L_2(\Omega)} \\
&= \|\Phi_\sigma(D_1) - \Phi_\sigma(D_2) + \Phi_\sigma(\psi(D_1)) \\
&\quad - \Phi_\sigma(\psi(D_2))\|_{L_2(\Omega)}.
\end{aligned}$$

Therefore, by following the proof of Reininghaus et al. (2015, Theorem 2) and by observing that $d_{W,1}(\psi(D_1),\psi(D_2)) = \|\psi(D_1)-\psi(D_2)\|_\infty$, the triangular inequality leads to

$$\|\Phi_\sigma^{\Psi_a}(D_1)-\Phi_\sigma^{\Psi_a}(D_2)\|_{L_2(\Omega)} \le \frac{1}{2\sqrt{\pi}\sigma}(\delta+\|\psi(D_1)-\psi(D_2)\|_\infty).$$

The proof is then concluded. □

Proposition 17.4.2 *Let* $\kappa_G(\cdot,\cdot,\kappa,\omega_{\text{arc}})$ *be the PWG kernel defined in* (17.4), *where* $\kappa(\boldsymbol{x},\boldsymbol{y}) = \exp(-\|\boldsymbol{x}-\boldsymbol{y}\|^2/(2\eta^2))$, *and let* D_1,D_2 *be r-dimensional PDs for some* $r\ge 0$. *By denoting as* $E_\kappa^{\Psi_a}(\mu_{D_1}^{\omega_{arc}}) = E_\kappa(\mu_{D_1\cup\psi(D_1)}^{\omega_{arc}})$, *we have*

$$\|E_\kappa^{\Psi_a}(\mu_{D_1}^{\omega_{arc}})-E_\kappa^{\Psi_a}(\mu_{D_2}^{\omega_{arc}})\| \le L d_B(D_1,D_2)+L^\Psi(D_1,D_2)$$

with

$$L^\Psi(D_1,D_2) = 2\omega_{arc}(\psi(D_1))\left(1-\exp\left(-\frac{\|\psi(D_1)-\psi(D_2)\|^2}{2\eta^2}\right)\right)+ \\ +|\omega_{arc}(\psi(D_1))-\omega_{arc}(\psi(D_2))|$$

and L as in (17.5).

Proof. We can follow the proof of Kusano et al. (2017, Theorem D.1) and the assumptions therein. As in Proposition 17.4.1, we impose the extension $\gamma(\psi(D_1)) = \psi(D_2)$ for the bijection between D_1 and D_2, obtaining then

$$\begin{aligned}
\|E_\kappa^{\Psi_a}(\mu_{D_1}^{\omega_{\text{arc}}})-E_\kappa^{\Psi_a}(\mu_{D_2}^{\omega_{\text{arc}}})\| \le &\sum_{\boldsymbol{x}\in D_1\cup\psi(D_1)} \omega_{\text{arc}}(\boldsymbol{x}))\|\kappa_G(\cdot,\boldsymbol{x})-\kappa_G(\cdot,\gamma(\boldsymbol{x}))\|_{\mathscr{F}} \\
&+\sum_{\boldsymbol{x}\in D_1\cup\psi(D_1)} |\omega_{\text{arc}}(\boldsymbol{x})-\omega_{\text{arc}}(\gamma(\boldsymbol{x}))|\|\kappa_G(\cdot,\gamma(\boldsymbol{x}))\|_{\mathscr{F}} \\
&+\sum_{\boldsymbol{y}\in(D_2\cup\psi(D_2))'} \omega_{\text{arc}}(\boldsymbol{y})\|\kappa_G(\cdot,\gamma(\boldsymbol{y}))\|_{\mathscr{F}} \\
&\le L d_B(D_1,D_2)+\omega_{\text{arc}}(\psi(D_1)))\|\kappa_G(\cdot,\psi(D_1))-\kappa_G(\cdot,\psi(D_2))\|_{\mathscr{F}} \\
&+|\omega_{\text{arc}}(\psi(D_1))-\omega_{\text{arc}}(\psi(D_2))|\|\kappa_G(\cdot,\psi(D_2))\|_{\mathscr{F}}.
\end{aligned}$$

We conclude by observing that

$$\|\kappa_G(\cdot,\psi(D_1))-\kappa_G(\cdot,\psi(D_2))\|_{\mathscr{F}} = 2\left(1-\exp\left(-\frac{\|\psi(D_1)-\psi(D_2)\|^2}{2\eta^2}\right)\right).$$

This concludes the proof. □

Proposition 17.4.3 *Let* κ_{SW} *be the SW kernel defined in* (17.6) *and let* D_1,D_2 *be r-dimensional PDs for some* $r\ge 0$. *We have*

$$d_{SW}^{\Psi_a}(D_1,D_2) \le d_{SW}(D_1,D_2)+d_{SW}(\psi(D_1),\psi(D_2)).$$

Proof. We follow the assumptions of Carrière et al. (2017, Theorems 3.3 and 3.4). Letting $\theta \in \mathbb{R}^2$, with $\|\theta\|_2 = 1$, we introduce $\mu_D^\theta = \sum_{p\in D} \delta_{\pi_\theta(p)}$ and $\mu_{D\mathscr{B}}^\theta = \sum_{p\in D} \delta_{\pi_\theta(p)\circ\pi_{\mathscr{B}}}$, where $\pi_\theta : \mathbb{R}^2 \to L(\theta)$ is the orthogonal projection onto the line $L(\theta) = \{\lambda\theta : \lambda \in \mathbb{R}\}$, $\pi_{\mathscr{B}}$ is the orthogonal projection onto the diagonal and δ is the Dirac function. Furthermore, let

$$\mathscr{W}(\mu,\nu) = \inf_{P\in\Pi(\mu,\nu)} \iint_{\mathbb{R}\times\mathbb{R}} |x-y| P(dx,dy)$$

where $\Pi(\mu,\nu)$ is the set of of measures on $\mathbb{R}^2$ with marginals μ and ν. Then,

and, under the condition of Carrière et al. (2017, remark 2.2), we have that

$$\mathscr{W}(\mu_{D_1}^\theta, \mu_{D_2}^\theta) = \|S_{D_1}^\theta - S_{D_2}^\theta\|_1,$$

where $S_{D_1}^\theta, S_{D_2}^\theta$ are the vectors of points $x \in D_1$ and $y \in D_2$, respectively, projected onto the line $L(\theta)$ and then sorted from the smallest value to the highest. Then, we have

$$\begin{aligned}
&\mathscr{W}(\mu_{\Psi_a(D_1)}^\theta + \mu_{\Psi_a(D_2)\mathscr{B}}^\theta, \mu_{\Psi_a(D_2)}^\theta + \mu_{\Psi_a(D_1)\mathscr{B}}^\theta) = \\
&\qquad = \|S_{\Psi_a(D_1)\cup\Psi_a(D_2)\mathscr{B}}^\theta - S_{\Psi_a(D_2)\cup\Psi_a(D_1)\mathscr{B}}^\theta\|_1 \\
&\qquad \le \|S_{D_1\cup D_2\mathscr{B}}^\theta - S_{D_2\cup D_1\mathscr{B}}^\theta\|_1 + \|S_{\psi(D_1)\cup\psi(D_2)\mathscr{B}}^\theta - S_{\psi(D_2)\cup\psi(D_1)\mathscr{B}}^\theta\|_1 \\
&\qquad = \mathscr{W}(\mu_{D_1}^\theta + \mu_{D_2\mathscr{B}}^\theta, \mu_{D_2}^\theta + \mu_{D_1\mathscr{B}}^\theta) \\
&\qquad + \mathscr{W}(\mu_{\psi(D_1)}^\theta + \mu_{\psi(D_2)\mathscr{B}}^\theta, \mu_{\psi(D_2)}^\theta + \mu_{\psi(D_1)\mathscr{B}}^\theta),
\end{aligned}$$

from which our statement follows. □

Corollary 17.4.1 *In the hypotheses of Proposition 17.4.3, we have*

$$d_{SW}^{\Psi_a}(D_1, D_2) \le 2\sqrt{2}(d_{W,1}(D_1,D_2) + \|\psi(D_1) - \psi(D_2)\|_\infty).$$

Proof. We can write

$$\begin{aligned}
d_{SW}^{\Psi_a}(D_1, D_2) &\le d_{SW}(D_1,D_2) + d_{SW}(\psi(D_1), \psi(D_2)) \\
&\le 2\sqrt{2}(d_{W,1}(D_1,D_2) + d_{W,1}(\psi(D_1), \psi(D_2))) \\
&\le 2\sqrt{2}(d_{W,1}(D_1,D_2) + \|\psi(D_1) - \psi(D_2)\|_\infty).
\end{aligned}$$

This concludes the proof. □

17.4.2 The Feature Extraction Map: Ψ_ρ

Let $\rho \in \mathbb{N}$ and $\widetilde{D_r}(\mathscr{X},\boldsymbol{\varepsilon}) = D_r(\mathscr{X},\boldsymbol{\varepsilon}) \setminus \mathscr{B}$. We define the set $\widetilde{D_r}(\mathscr{X},\boldsymbol{\varepsilon},\rho)$, which consists of the ρ most persistent elements in $\widetilde{D_r}(\mathscr{X},\boldsymbol{\varepsilon})$, and the function

$$\Psi_\rho(D_r(\mathscr{X},\boldsymbol{\varepsilon})) = \widetilde{D_r}(\mathscr{X},\boldsymbol{\varepsilon},\rho) \cup \psi\big(D_r(\mathscr{X},\boldsymbol{\varepsilon}) \setminus \widetilde{D_r}(\mathscr{X},\boldsymbol{\varepsilon},\rho)\big) \cup \mathscr{B}.$$

We remark that $(D_r(\mathscr{X},\boldsymbol{\varepsilon}) \setminus \widetilde{D_r}(\mathscr{X},\boldsymbol{\varepsilon},\rho)) \in \mathfrak{D}_r(\boldsymbol{\varepsilon})$, therefore $\Psi_\rho(D_r(\mathscr{X},\boldsymbol{\varepsilon}))$ is well defined.

The map Ψ_ρ performs a *feature extraction* procedure. Indeed, the resulting persistence diagram consists of the most ρ persistent elements, $\mathscr{B}$, while the remaining elements, which are possibly large in numbers, are *compressed* into a single generator. Nevertheless, it can be related to Ψ_a by the formula

$$\Psi_\rho = \widetilde{D_r}(\mathscr{X},\boldsymbol{\varepsilon},\rho) \cup \Psi_a(D_r(\mathscr{X},\boldsymbol{\varepsilon}) \setminus \widetilde{D_r}(\mathscr{X},\boldsymbol{\varepsilon},\rho)) \setminus (D_r(\mathscr{X},\boldsymbol{\varepsilon}) \setminus \widetilde{D_r}(\mathscr{X},\boldsymbol{\varepsilon},\rho)).$$

As we experiment in Section 17.5, Ψ_ρ proves especially useful when many generators are associated with *noisy* structures in the data, resulting in significant savings in computational cost.

17.4.3 The Auxiliary Function ψ

We define the auxiliary function $\psi : \mathfrak{D}_r(\boldsymbol{\varepsilon}) \to \mathbb{R}^2_{\geq \mathscr{B}}$ as

$$\psi(D_r(\mathscr{X},\boldsymbol{\varepsilon})) = \frac{1}{W} \sum_{\boldsymbol{x} \in \widetilde{D_r}(\mathscr{X},\boldsymbol{\varepsilon})} w(\boldsymbol{x})\boldsymbol{x} \tag{17.7}$$

where $w : \mathbb{R}^2_+ \longrightarrow \mathbb{R}_+$ is a weight function and $W = \sum_{\boldsymbol{x} \in \widetilde{D_r}(\mathscr{X},\boldsymbol{\varepsilon})} w(\boldsymbol{x})$. We observe that $\widetilde{D_r}(\mathscr{X},\boldsymbol{\varepsilon})$ contains a finite number of elements (generators), therefore the sum in (17.7) is always defined (see also Remark 17.4.1).

We can consider the following weights.

1. Let $w_1(\boldsymbol{x}) = 1/M$, being $M = |\widetilde{D_r}(\mathscr{X},\boldsymbol{\varepsilon})|$, that is, the cardinality of the multiset, in which each element is counted with its multiplicity. We denote as *center of uniform mass* the resulting auxiliary function

$$\psi_1(D_r(\mathscr{X},\boldsymbol{\varepsilon})) = \frac{1}{M} \sum_{\boldsymbol{x} \in \widetilde{D_r}(\mathscr{X},\boldsymbol{\varepsilon})} \boldsymbol{x}.$$

2. Let $\boldsymbol{x} = (b,d) \in \widetilde{D_r}(\mathscr{X},\boldsymbol{\varepsilon})$, where $b,d \in \mathbb{R}_+$ are the birth–death *time* of the element $\boldsymbol{x}$ (see (17.2)), and let $w_2(x) = d - b$ be the persistence of $\boldsymbol{x}$. We denote as *center of persistence* the auxiliary function. Letting $C =$

$$\sum_{\mathbf{x}=(b,d)\in\widetilde{D_r}(\mathscr{X},\boldsymbol{\varepsilon})} (d-b)$$

$$\psi_2(D_r(\mathscr{X},\boldsymbol{\varepsilon})) = \frac{1}{C} \sum_{\mathbf{x}=(b,d)\in\widetilde{D_r}(\mathscr{X},\boldsymbol{\varepsilon})} (d-b)\mathbf{x}.$$

Both options aim to extract a *representative* element from the generators of the persistence diagram. While the center of uniform mass is the *barycentre* of the elements of the multiset, the center of persistence assigns different weights according to the persistence of the elements. This is a natural choice for analysis, since elements with low persistence are more likely to be related to *noise* structures that give rise to the filtration. In contrast, aspects of significant persistence are associated with more representative geometric features of the dataset (see the example in Figure 17.4). For these reasons, in Section 17.5 we mainly use the function ψ_2 in our experiments, but we also compare ψ_1 and ψ_2 in Section 17.5.1.

Remark 17.4.1 *If we take* $\mathbf{x} \in D_r(\mathscr{X},\boldsymbol{\varepsilon})$ *in* (17.7), *additional conditions on the weight function* w *are needed to guarantee the convergence of the sum. However, such an infinite setting is not meaningful to analyze, since elements in the bisector carry no topological information about the dataset. As a further observation, the center of persistence might be formally computed by summing over* $D_r(\mathscr{X},\boldsymbol{\varepsilon})$, *as in this case* $w(\mathbf{x}) = 0$ *for* $\mathbf{x} \in \mathscr{B}$.

Remark 17.4.2 *Since* $D_r(\mathscr{X},\boldsymbol{\varepsilon})$ *contains an infinite number of elements for all* $\mathscr{X} \subset \Omega$ *by definition, the VSPK* κ_Ψ *is still well defined on* $\mathfrak{D}_r(\boldsymbol{\varepsilon}) \times \mathfrak{D}_r(\boldsymbol{\varepsilon})$. *Moreover, if* κ *is (strictly) positive definite, so it is* κ_Ψ.

Remark 17.4.3 *Referring to Remark 17.3.1, if* κ *is a PWG or SW kernel, then* κ_Ψ *can be directly expressed in terms of the distance* $d_\Psi(\cdot,\cdot)$ *induced in the variably scaled setting.*

17.5 Experiments

We aim to show how the SVM classifier may benefit from the introduced VSPKs. In all experiments we present, the kernels are handled using Python 3.8 and the `scikit-learn` module (Pedegrosa, F. et al., 2011). Scikit-learn is a Python module that integrates different machine learning algorithms for medium-scale supervised and unsupervised problems. The package has been designed as a machine learning tool for non-specialists, implemented in a general-purpose high-level language like Python.

Persistence diagrams are constructed via the Python modules

- *persim* `https://persim.scikit-tda.org/en/latest/` and *ripser* `https://ripser.scikit-tda.org/en/latest/`: both developed within the project `scikit-learn`;
- *GUDHI* `https://gudhi.inria.fr/`.

To reproduce the following experiments, the open-source Python code is available at

`https://github.com/reevost/vspk_code`.

We validate the following hyperparameters. About the considered kernels, we follow the guidelines provided by the authors of the seminal papers.

- Concerning the SVM classifier, we validate $\zeta \in \{10^j \mid j = -3, \ldots, 3\}$.
- Concerning the PSS kernel, we take $\sigma \in \{10^j \mid j = -3, \ldots, 3\} \cup \{5 \cdot 10^j \mid j = -3, \ldots, 2\}$.
- Concerning the PWG kernel, the parameters C and τ of the PWGK are chosen in $\{10^j \mid j = -2, \ldots, 2\}$, while δ is set to 10 (see Kusano et al., 2017, Theorem 3.2). Moreover, as the underlying standard kernel, we use the Gaussian.
- Concerning the SW kernel, σ is obtained following the procedure carried out in (Carrière et al., 2017, §4).

To assess the performance of the binary classifiers, we consider the following scores. Letting TP=true positives, TN=true negatives, FP=false positives, and FN=false negatives, we define using

- Accuracy score

$$\text{Accuracy} = \frac{\text{TP+TN}}{\text{TP+TN+FP+FN}}.$$

- f_1-score (cf. e.g. van Rijsbergen, 1979 and Bandiziol and De Marchi, 2024)

$$f_1\text{-score} = 2 \cdot \frac{\text{precision} \cdot \text{recall}}{\text{precision} + \text{recall}},$$

where

$$\text{Precision} = \frac{\text{TP}}{\text{TP+FP}}, \quad \text{Recall} = \frac{\text{TP}}{\text{TP} + \text{FN}}.$$

The f_1-score is indeed the harmonic mean of precision and recall, and it is then widely used in literature because it provides a useful trade-off between these two important metrics.

It is worth mentioning that in the multiclass scenario (i.e., n classes) other metrics are used (cf. Bandiziol, 2025). For example, the `Balanced_Accuracy` that considers the number of correct predictions per class, called again *Recall*, and then takes the average. More precisely,

$$\text{Recall}_i = \frac{\text{samples of the } i-th \text{ class correctly classified}}{\text{all test samples of the } i-th \text{ class}}$$

so that

$$\text{Balanced\_Accuracy} = \frac{\sum_{i=1}^{n} Recall_i}{n}.$$

17.5.1 Alzheimer's Disease Diagnosis

The Open Access Series of Imaging Studies (OASIS) is a project aimed at making neuroimaging data sets freely available to the scientific community. In particular, OASIS-3 is a compilation of MRI and PET imaging and related clinical data from 1,098 participants collected across several ongoing studies at the Washington University Knight Alzheimer's Disease Research Center over 15 years. Imaging data are accompanied by dementia status, APOE status, and longitudinal clinical and cognitive outcomes (LaMontagne et al., 2019; Pachauri et al., 2011).

We consider a subset of the study group to ensure balanced data.

A summary of demographic and neuropsychological details of the subjects considered in our study is presented in Table 17.1.

Table 17.1 Demographic details and baseline cognitive status measures of the study population. M=Mean, S.D.=Standard Deviation. In Hand Preferences, A=All, L=Left, R=Right. Finally, the MMSE (Mini-Mental State Examination) is the scoring system for this neuropsychological test. A score less than 23 indicates progressive cognitive decline

	AD (M)	AD (S.D.)	Control (M)	Control (S.D.)
Subjects	225	-	248	-
Gender (F/M)	114/111	-	126/122	-
Hand Pref. (A/L/R)	5/23/197	-	6/26/216	-
Age	74.41	7.60	65.21	9.62
Education years	14.77	3.08	16.04	2.51
MMSE	20.33	6.38	29.27	1.30

For each subject, we construct the PDs using estimates of cortical thickness at 34 points in both hemispheres, for a total of 64 values. For simplicity, in the study, we use the exact coordinates for the above-mentioned points across all subjects. The coordinates are computed with the *scipy* toolbox (Virtanen et al., 2020). From these coordinates, we construct the PDs and extract 1D and 2D topological features, that is, the generators of H_1 and H_2.

In Figure 17.5, we show two examples of PDs, and in Figure 17.6 we highlight the generator added as the center of persistence.

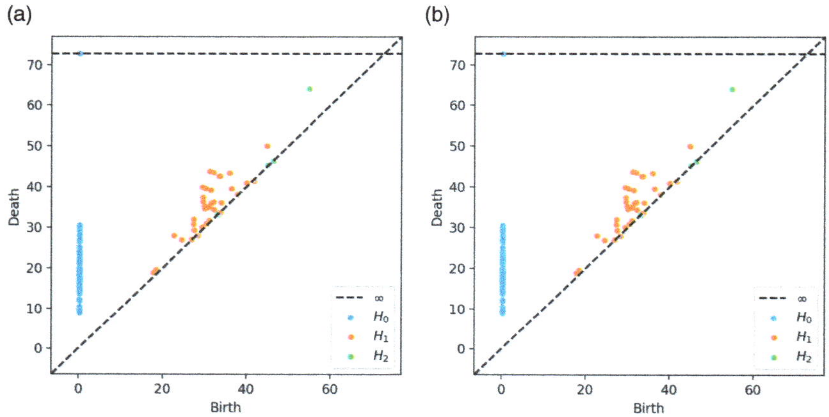

Figure 17.5 Persistence diagrams of an AD subject with a MMSE of 30 (a) and the persistence diagram of a control subject with a MMSE of 7 (b).

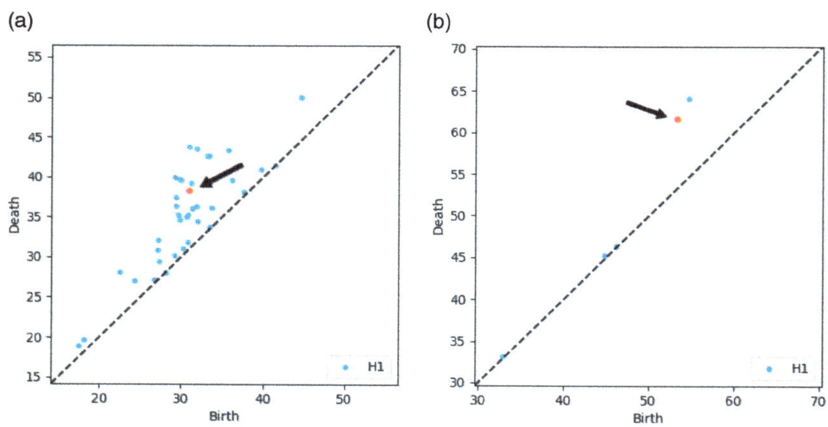

Figure 17.6 1-dimensional (a) and 2D (b) persistence diagram of an AD subject. The red dot, pointed by a black arrow, is the added *center of persistence* via Ψ_a and ψ_2.

We evaluate the performance of an SVM classifier using the presented PSS, PWG, and SW kernels, both in the classical and variably scaled settings.

In each test, we perform a random 70%/30% split of the dataset for training and testing, and we use five-fold CV on the training set to tune the hyperparameters. The results displayed in Tables 17.2–17.4 have been averaged over 10 runs of tests. There, the reported validation time refers to the time required by the whole CV process.

Table 17.2 OASIS-3 dataset. Results of SVM classification obtained using H_1 and H_2 persistence diagrams and the SW kernel

	Ψ	ψ	Accuracy	f_1-score	Validation time (s)
SW (H_1)	-	-	0.741	0.716	220
VSP-SW (H_1)	Ψ_a	ψ_2	0.732	0.700	223
VSP-SW (H_1)	Ψ_ρ	ψ_2	0.731	0.693	184
SW (H_2)	-	-	0.741	0.712	183
VSP-SW (H_2)	Ψ_a	ψ_2	0.753	0.720	300

Table 17.3 OASIS-3 dataset. Results of SVM classification obtained using H_1 and H_2 persistence diagrams and the PWG kernel

	Ψ	ψ	Accuracy	f_1-score	Validation time (s)
PWGK (H_1)	-	-	0.749	0.723	18165
VSP-PWG (H_1)	Ψ_a	ψ_2	0.750	0.726	18082
VSP-PWG (H_1)	Ψ_ρ	ψ_2	0.759	0.735	8731
PWG (H_2)	-	-	0.716	0.699	4422
VSP-PWG (H_2)	Ψ_a	ψ_2	0.709	0.683	4713

Table 17.4 OASIS-3 dataset. Results of SVM classification obtained using H_1 and H_2 persistence diagrams and the PSS kernel

	Ψ	ψ	Accuracy	f_1-score	Validation time (s)
PSS (H_1)	-	-	0.743	0.721	9238
VSP-PSS (H_1)	Ψ_a	ψ_2	0.752	0.728	9045
VSP-PSS (H_1)	Ψ_ρ	ψ_2	0.750	0.723	3330
PSS (H_2)	-	-	0.781	0.762	2825
VSP-PSS (H_2)	Ψ_a	ψ_2	0.775	0.755	3120

For the variably scaled setting, we consider Ψ_a for both H_1 and H_2 diagrams, while Ψ_ρ, with $\rho = 10$, is employed with H_1 only, since H_2 diagrams are

Table 17.5 OASIS-3 dataset. Results of SVMs classification obtained by using the center of mass and persistence alone in place of the persistence diagrams

	Accuracy	f_1-score
ψ_1	0.56	0.55
ψ_2	0.65	0.72

limited in the number of generators, and therefore compressing features is not meaningful.

Furthermore, we use ψ_2 as an auxiliary function. Indeed, as highlighted in Table 17.5, ψ_2 definitely outperforms ψ_1 in our setting. Moreover, we observe that the performance achieved by the auxiliary function alone, that is, using the centers of mass or persistence in place of the persistence diagram, is definitely not competitive with the classical and variably scaled settings.

We observe that VSPKs are competitive with the classical setting, improving performance in some cases. On the other hand, using Ψ_ρ yields consistent savings in validation time.

17.5.2 Orbit Recognition

As the second experiment, we follow the idea proposed in Adams et al. (2017) and we analyze the linked twisted map, which models fluid flows. The corresponding orbits are computed via the discrete system

$$\begin{cases} x_{n+1} = x_n + r y_n (1 - y_n) & \mod 1 \\ y_{n+1} = y_n + r x_{n+1} (1 - x_{n+1}) & \mod 1 \end{cases}$$

where $(x_0, y_0) \in [0,1] \times [0,1]$ is the initial position and $r > 0$ is a real parameter that influences the orbit. The topological structure of the orbit changes with the initial position and r, as displayed in Figures 17.7 and 17.8, where we depict the first 1,000 *iterations* $\{(x_n, y_n) : n = 0, \ldots, 1000\}$.

According to Adams et al. (2017), we construct the PDs as follows:

- choose a set parameters r as classification labels (we took $r = \{2.5, 3.5, 4, 4.1, 4.3\}$;
- for each label, we compute the first 1,000 points of 50 orbits, with a random starting point, giving a dataset consisting of 250 elements;
- compute the persistence diagram related to each orbit.

Here, since each persistence diagram has a considerable number of generators ($\approx 10^5$), we restrict to the most $\rho = 10$ persistent elements in each diagram.

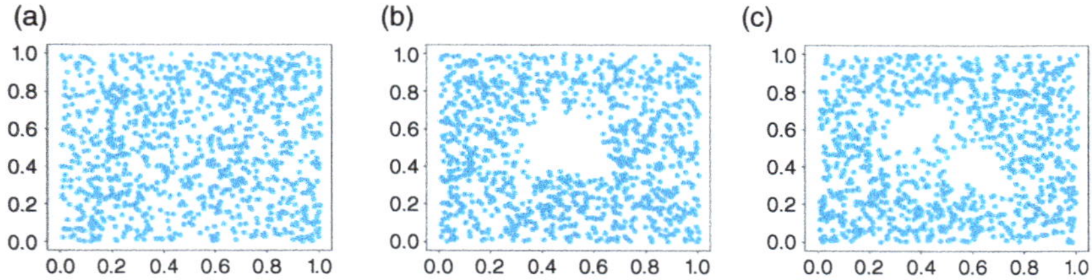

Figure 17.7 Fixed $(x_0, y_0) \in [0,1]^2$, the orbits resulting from the linked twisted map taking $r = 2.5, 4.1$, and 4.3, (a), (b), and (c), respectively.

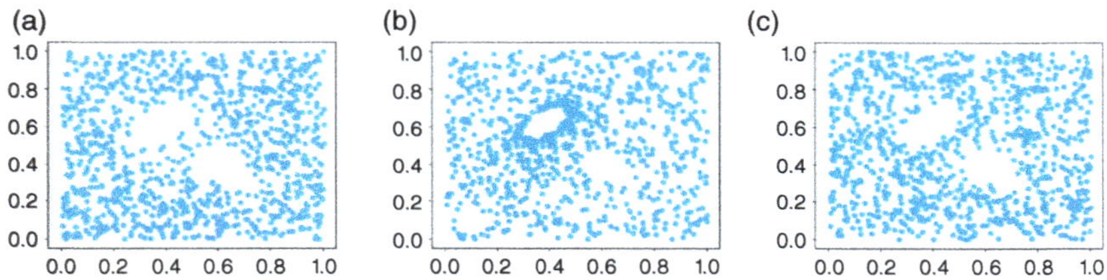

Figure 17.8 (a–c) Fixed $r = 4.3$, the orbits resulting from the linked twisted map taking different starting points $(x_0, y_0) \in [0,1]^2$.

Indeed, in our experiments, we observed that increasing the number of generators does not significantly improve classification performance but does increase computational cost. More precisely, here Ψ_a is computed with respect to such 10 elements, while in the case of Ψ_ρ the discarded less persistent generators are compressed in a unique element via ψ_2. Moreover, since we are dealing with 2D orbits, we compute only the H_1 homology group.

As in Section 17.5.1, we consider a five-fold CV on the training set, and the results displayed in Tables 17.6, 17.7, and 17.8 are averaged over 10 runs with a 70%/30% training-test split of the data.

Table 17.6 Orbit recognition. Results of SVMs classification on H_1 persistence diagrams using the SW kernel

	Ψ	ψ	Accuracy	f_1-score
SW (H_1)	-	-	0.832	0.831
VSP–SW (H_1)	Ψ_a	ψ_2	0.814	0.812
VSP–SW (H_1)	Ψ_ρ	ψ_2	0.833	0.832

Table 17.7 Orbit recognition. Results of SVMs classification on H_1 persistence diagrams using the PWG kernel

	Ψ	ψ	Accuracy	f_1-score
PWG (H_1)	-	-	0.858	0.866
VSP-PWG (H_1)	Ψ_a	ψ_2	0.858	0.866
VSP-PWG (H_1)	Ψ_ρ	ψ_2	0.846	0.853

Table 17.8 Orbit recognition. Results of SVMs classification on H_1 persistence diagrams using the PSS kernel

	Ψ	ψ	Accuracy	f_1-score
PSS (H_1)	-	-	0.806	0.803
VSP-PSS (H_1)	Ψ_a	ψ_2	0.824	0.821
VSP-PSS (H_1)	Ψ_ρ	ψ_2	0.826	0.823

17.5.3 3D Shape Segmentation

Following the experiment proposed also in Carrière et al. (2017). We consider some categories of the mesh segmentation benchmark introduced in Chen et al. (2009), which contains different 3D shapes of several categories. Precisely, we consider four different categories/datasets: *Ant*, *Airplane*, *Bird* and *Octopus*. In the datasets, each sample is a *face*, which is represented as a triplet of 3D points. For each category, the classification task is to associate each sample with its corresponding segmentation label, which describes the sample's relative position with respect to the shape. We note that four classification tasks are considered, and each dataset has a different number of labels, as detailed in Table 17.9. A H_1 persistence diagram is then computed for each face by using the geodesic distance on the 3D shape; for more details concerning the datasets, we refer to Carrière et al. (2015), while in Figure 17.9 we display some examples of shape segmentation.

Table 17.9 Details for the considered datasets

Category	Number of samples	Number/type of labels
Ant	72546	Five (head, torso, tail, legs, antennas)
Airplane	59,546	Four (head, torso, tail, wings)
Bird	50,902	Four (head, torso, tail, wings)
Octopus	66248	Two (head, legs)

As in the Sections 17.5.1 and 17.5.2, we perform five-fold CV on the training set, and the results displayed in Tables 17.10 and 17.11 are averaged over 10 runs with a 20%/80% training-test split.

We remark that the elements of the PDs in this experiment are limited in numbers and free of noisy generators, because of the particular construction of

Table 17.10 Accuracy achieved by the SVM classifiers in the carried out tests. In the variably scaled setting, we used Ψ_a and ψ_2 functions

	PSS (H_1)	VSP–PSS (H_1)	SW (H_1)	VSP–SW (H_1)
Ant	0.925	0.920	0.925	0.934
Airplane	0.822	0.820	0.822	0.843
Bird	0.881	0.858	0.881	0.882
Octopus	0.915	0.925	0.915	0.942

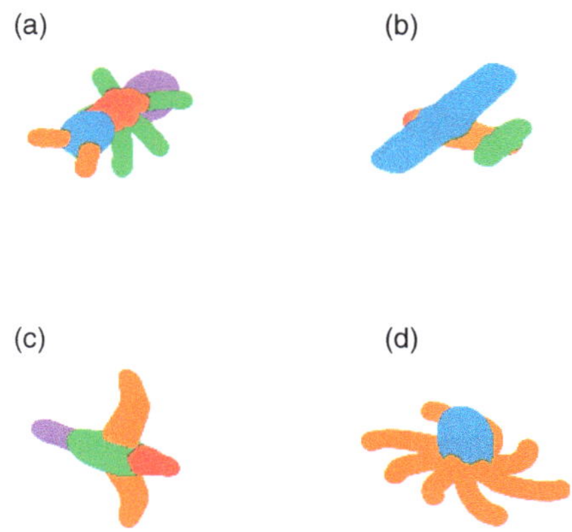

Figure 17.9 Examples of 3D shapes and corresponding segmentation labels. (a) Ant, (b) Airplane, (c) Bird, and (d) Octopus.

Table 17.11 f1-score achieved by the SVM classifiers in the carried out tests. In the variably scaled setting, we used Ψ_a and ψ_2 functions

	PSS (H_1)	VSP–PSS (H_1)	SW (H_1)	VSP–SW (H_1)
Ant	0.910	0.905	0.910	0.913
Airplane	0.811	0.813	0.811	0.814
Bird	0.826	0.779	0.826	0.807
Octopus	0.912	0.922	0.912	0.940

the diagrams obtained via the geodesic metric. Therefore, the usage of Ψ_ρ is not significant.

17.6 Conclusion and Future Developments

The classification of PDs using SVM enabled the introduction of a suitable definition of kernels acting on a pair of PDs. We analyzed the behavior of four *persistence kernels* (PKs) and their *scaled* version, making experiments for tuning the proper hyperparameters. This comparative analysis was largely absent in the literature but has recently been conducted in De Marchi et al. (2022) and Bandiziol and De Marchi (2024), leading to the conclusion that SWK provides the best choice among the tested kernels, mainly for two reasons.

First, it predicts labels better than the other. It is computationally less expensive, since its `preGram` matrix depends on no parameters and can be computed only once at the beginning of the code. In Bandiziol and De Marchi (2024), other datasets were considered, and it remains an open question why, in the ENZYMES dataset, all kernels perform poorly, resulting in lower accuracy than in other datasets.

Another interesting application of persistent homology that is worth mentioning is the estimation of the *Intrinsic Dimension* (ID) of a manifold or, more generally, a point cloud. Indeed, in the framework of Machine Learning, the user typically works with large amounts of data and a considerable number of features. The number of the *essential features* is precisely what the Intrinsic Dimension is about. In the PhD dissertation by Cinzia Bandiziol (Bandiziol, 2025, Chapter 5), the most popular methods for estimating ID today have been discussed, among them the *box-counting dimension* and the *correlation dimension*. We point out how PH can be used to create ID estimators, describing *i-Dimensional Persistent Homology Fractal Dimension*, *Persistent Homology Dimension*, and *Persistent Homology Complexity*. The numerical tests done in the dissertation reveal how *i-Dimensional Persistent Homology Fractal Dimension* ends up being "optimal" in the sense that its α generalization obtains its best value when $\alpha = 1$ Bandiziol et al. (2025). Further analysis is definitely needed to understand better how the estimator poorly approximates the ID of some datasets, to improve, in the future, the methods themselves.

These final considerations show that TDA by persistent homology is an active area of research, combining pure mathematics with data science. Data analysis is the present and future of our society. We must make more bridges between mathematics and the applied sciences.

18
Software

This chapter provides a comprehensive overview of the software packages developed by the authors over the past decades, either as primary developers or as co-authors of the respective projects. All software is publicly available and released under Creative Commons or similar open-source licenses. For specific license details, readers are encouraged to consult the information provided in the links mentioned in this chapter.

The programming languages employed across the repositories vary depending on the intended research community and computational goals. In particular, MATLAB remains the dominant choice within the traditional approximation and scientific computing communities, owing to its extensive numerical libraries and ease of prototyping. In contrast, PYTHON has gained significant popularity in machine learning and data-driven research, providing a broader user base, better integration with modern frameworks, and enhanced accessibility for algorithm deployment.

For foundational techniques (see Chapters 1–4) and classical kernel methods, readers are referred to the comprehensive MATLAB codes collected in Fasshauer (2007) and Fasshauer and McCourt (2015). In this chapter, we present additional software packages or modules developed by the authors, highlighting their functionality, implementation details, and relevant usage examples.

Importantly, all the software presented in this chapter reproduces the experiments, figures, and numerical results discussed throughout this book, allowing readers to verify, explore, and extend the examples hands-on.

18.1 Software for *Stable Kernels*

This section gathers software tools specifically designed to enhance the numerical stability and accuracy of kernel-based methods. These packages implement the algorithms as discussed in the Chapters 5 and 6.

ERA Algorithm. Selecting appropriate shape parameters is crucial for the stability and accuracy of kernel-based approximations. The ERA algorithm, discussed in Chapter 5, represents an efficient extension of the classical Rippa scheme to general k-fold CV scenarios. The ERA package provides a fully functional MATLAB implementation, freely available at:

`https://github.com/cesc14/RippaExtCV`

In addition to the deterministic version, the repository includes a stochastic variant, the SERA algorithm, that introduces probabilistic node selection to improve robustness. The software was developed by Francesco Marchetti, who provides a `demo.m` file to facilitate understanding and practical usage of the algorithms. The package allows users to experiment with different CV strategies and kernel types.

Citation policy. If you use this software in your research or publications, please acknowledge the authors by citing
F. Marchetti, The extension of Rippa's algorithm beyond LOOCV, Applied Mathematics Letters, 120, 107262, 2021.
L. Ling, F. Marchetti, A stochastic extended Rippa's algorithm for LpOCV, Applied Mathematics Letters, 129, 2022.

Rational Kernels. An alternative approach for achieving stability in kernel interpolants is provided by rational kernels, as described in Chapter 6. The associated MATLAB implementation includes an efficient PU-accelerated implementation for fast computations. The software is available at:

`https://github.com/emmaA89/FastRVSK_PU`

This software enables users to recreate every numerical example presented in the chapter on rational kernels, including stability comparisons with classical kernels.

The authors of this package (S. De Marchi, A. Martínez, and E. Perracchione) provide a `demo.m` file to illustrate the main functionalities, enabling users to understand the workflow of rational kernel interpolation quickly. This tool is handy for applications requiring stable approximations with large datasets, where classical RBF methods may suffer from numerical instability.

Citation policy. If you use this software in your research or publications, please acknowledge the authors by citing
S. De Marchi, A. Martínez, and E. Perracchione, *Fast and stable rational RBF-based partition of unity interpolation*, Journal of Computational and Applied Mathematics, 349, 331–343, 2019.

18.2 Software for Local and Efficient Methods

This section collects all software implementations devoted to local approximation strategies and efficient tools. The presented codes closely follow the methods described in the book (PU, LGSVM, and ERBA) and provide reference demos. These tools serve as practical gateways to exploring the interplay between locality, partitioning, and efficient kernel-based approximation.

PU scheme. The PU method is central to many local approximation and meshfree strategies. Several implementations exist in the literature; among them, the library developed by Cavoretto et al. (2022) provides both MATLAB and PYTHON versions, available at:

`https://github.com/sandro-lancellotti/PU`

This library is based on the efficient partitioning data structures described in Chapter 7, allowing scalable computations even for large datasets. Users can choose between MATLAB (`usage_example.m`) and PYTHON (`usage_example.py`) implementations. This implementation allows readers to explore the influence of the covering strategy, the efficiency of partitioning structures, and the effect of local accuracy.

Citation policy. If you use this software in your research or publications, please acknowledge the authors by citing
R. Cavoretto, A. De Rossi, S. Lancellotti, E. Perracchione, Software Implementation of the Partition of Unity Method, Dolomites Research Notes on Approximation, 15(2), 2022.

LGSVM method. The LGSVM framework combines multiple local SVM models to construct a global classifier. The PYTHON implementation, discussed in Chapter 14, is available at:

`https://github.com/cesc14/LGSVM`

The scikit-learn-compatible estimator supports `fit`, `predict`, and `fit_predict` methods, and the repository includes a demo script (`demo_lgsvm.py`) illustrating its use on benchmark datasets. Local-to-global SVMs are particularly advantageous for large-scale classification problems, where data locality can be exploited to improve computational efficiency and accuracy. This makes the LGSVM repository a valuable tool for exploring locality-driven machine learning strategies.

Citation policy. If you use this software in your research or publications, please acknowledge the authors by citing
F. Marchetti, E. Perracchione, *Local-to-Global Support Vector Machines (LGSVMs)*, Pattern Recognition, 132, 108920, 2022.

ERBA scheme. ERBA provides a complementary approach to PU-based methods, as detailed in Chapter 8. Efficient reduced basis algorithm iteratively removes nodes from a full dataset according to a selection criterion, constructing a surrogate model that balances accuracy and sparsity. The MATLAB implementation is freely available at:

`https://github.com/cesc14/ERBA`

The package, developed by F. Marchetti and E. Perracchione, includes demo files to illustrate typical workflows. Efficient reduced basis algorithm is particularly useful for large-scale kernel approximation problems, offering a systematic strategy for dimensionality reduction while preserving essential information.

This package is therefore indispensable for those wishing to understand and experiment with strategies for kernel-based sparsification.

Citation policy. If you use this software in your research or publications, please acknowledge the authors by citing
F. Marchetti, E. Perracchione, *Efficient Reduced Basis Algorithm (ERBA) for kernel-based approximation*, Journal of Scientific Computing, 91, 2022, 41.

18.3 Software for *Optimal* Kernels

This section contains the software dedicated to data-driven and optimization-based kernel design. From kernel metric learning to feature reduction, providing a practical platform for exploring kernel optimization techniques.

2L-KOGA scheme. The 2L-KOGA algorithm (Chapter 15) optimizes kernels in a data-driven manner and subsequently performs greedy node selection using VKOGA.

The PYTHON software is available at:

`https://gitlab.mathematik.uni-stuttgart.de/pub/ians-anm/paper-2023-data-driven-kernel-designs`

This implementation provides a powerful environment for experimenting with optimized kernels in both synthetic and real-world scenarios.

Developed by Tizian Wenzel, Francesco Marchetti, and Emma Perracchione, the repository includes multiple demo scripts reproducing the results of Wenzel et al. (2024a). 2L-KOGA is particularly suited for learning high-dimensional kernels with structured sparsity and efficient greedy selection.

Citation policy. If you use this software in your research or publications, please acknowledge the authors by citing
T. Wenzel, F. Marchetti, E. Perracchione, *Data-Driven Kernel Designs for*

Optimized Greedy Schemes: A Machine Learning Perspective, SIAM Journal of Scientific Computing, 46(1), C101–C126, 2024.

2L-FUSE algorithm. Building on kernel optimization, the 2L-FUSE algorithm (Chapter 15) enables feature reduction by learning a data-adaptive kernel metric. Eigen-decomposition identifies optimal directions for sparsity enhancement. The software is available at:

`https://github.com/fabianacamattari/2L-FUSE`

A demo is provided via `main.py` or `main.ipynb`, making it easy to reproduce the examples and understand the workflow. The demos guide users through the same workflows illustrated in Chapter 15, making it easy to know how kernel metric learning leads to effective dimension reduction.

Citation policy. If you use this software in your research or publications, please acknowledge the authors by citing
F. Camattari, S. Guastavino, F. Marchetti, E. Perracchione, *Feature Understanding and Sparsity Enhancement via 2-Layered Kernel Machines (2L-FUSE)*, preprint, 2025.

18.4 Software VSKs

This section presents software for VSKs, with applications to feature augmentation and adaptive scaling. Each repository provides demos allowing exploration of the behavior of VSK models under different scaling functions and learning strategies.

VS(D)Ks. A basic MATLAB implementation of VSKs, focusing on discontinuous test cases, is available at:

`https://github.com/emmaA89/VSDKs`

The demo file `Demo.m`, authored by S. De Marchi, F. Marchetti, and E. Perracchione, provides a practical example for testing and understanding the methodology. Such demo script mirrors the tests shown in the book, including reconstruction accuracy, artefact reduction near jumps, and comparisons with classical methods. This package provides a direct way to visualize how VSKs adapt to spatial variability, supporting a deeper understanding of the theory presented in Chapter 9.

Citation policy. If you use this software in your research or publications, please acknowledge the authors by citing
S. De Marchi, F. Marchetti, E. Perracchione, *Jumping with Variably Scaled*

Discontinuous Kernels (VSDKs), BIT Numerical Mathematics, 60(2), 441–463, 2020.

SVM–VSKS. The SVM–VSK package extends VSKs to SVMs. The free PYTHON software is available at:

`https://github.com/emmaA89/SVM-VSK`

Developed by C. Campi, F. Marchetti, and E. Perracchione, the repository includes demo scripts for both toy models (`TestToyModel.py`) and real datasets (`TestRealCancerData.py`), illustrating feature augmentation and alternative feature extraction approaches. The demos enable users to explore how variably scaled kernels might enhance SVM performance.

Citation policy. If you use this software in your research or publications, please acknowledge the authors by citing
C. Campi, F. Marchetti, E. Perracchione, *Learning via variably scaled kernels*, Advances in Computational Mathematics, 47(51), 2021.

δNN–VSKs. The δNN–VSK algorithm (Chapter 16) offers a user-independent, adaptive solution for learning the scaling function of VSKs. The PYTHON implementation is available at:

`https://github.com/Fra0013To/VSKlearning`

The software, developed by G. Audone, F. Della Santa, E. Perracchione, and S. Pieraccini, includes demos for both continuous and discontinuous kernels (`example_deltaNN_VSK.py`). This software is essential for exploring advanced and automated scaling strategies in kernel methods.

Citation policy. If you use this software in your research or publications, please acknowledge the authors by citing
G. Audone, F. Della Santa, E. Perracchione, S. Pieraccini, *A recipe based on Lebesgue functions for learning Variably Scaled Kernels via Discontinuous Neural Networks*, Journal of Computational and Applied Mathematics, 469, 116653, 2025.

18.5 Software for Persistence Kernels

In the package discussed in this section, we interpret VSKs in the framework of persistent homology, whose outputs, persistence diagrams, are considered in supervised learning approaches. We call them VSPKs. The free PYTHON software is available at:

`https://github.com/reevost/vspk_code`

Citation policy. If you use this software in your research or publications, please acknowledge the authors by citing

P. J. LaMontagne, T. L. Benzinger, J. C. Morris, S. Keefe, R. Hornbeck, C. Xiong, E. Grant, J. Hassenstab, K. Moulder, A. Vlassenko, et al., Oasis-3: longitudinal neuroimaging, clinical, and cognitive dataset for normal aging and alzheimer disease, MedRxiv (2019).
M. Carriere, M. Cuturi, S. Oudot, Sliced Wasserstein kernel for persistence diagrams, in: International Conference on Machine Learning, PMLR, 2017, pp. 664–673.
G. Kusano, K. Fukumizu, Y. Hiraoka, Kernel method for persistence diagrams via kernel embedding and weight factor, The Journal of Machine Learning Research 18 (1) (2017) 6947–6987.
J. Reininghaus, S. Huber, U. Bauer, R. Kwitt, A stable multi-scale kernel for topological machine learning, in: Proceedings of the IEEE conference on computer vision and pattern recognition, 2015, pp. 4741–4748.

18.6 A Note on the Software for Astronomical Imaging

The software used to create the experiments in Chapter 11 for the already dismissed RHESSI telescope is written in IDL (in the SolarSoftWare [SSW] tree Schwartz et al., 2002). People of the kernel community is usually not familiar with that language. Fortunately, we are working together with the MIDA lab of the University of Genova on a PYTHON implementation of the interpolation/extrapolation algorithm presented in this book. We already have a free available implementation at

`https://github.com/MatteTro/GreedyIP`

It works for simulated data, precisely for the synthetic configuration plotted in Figure 18.1. The library will be updated for real data in a near future.

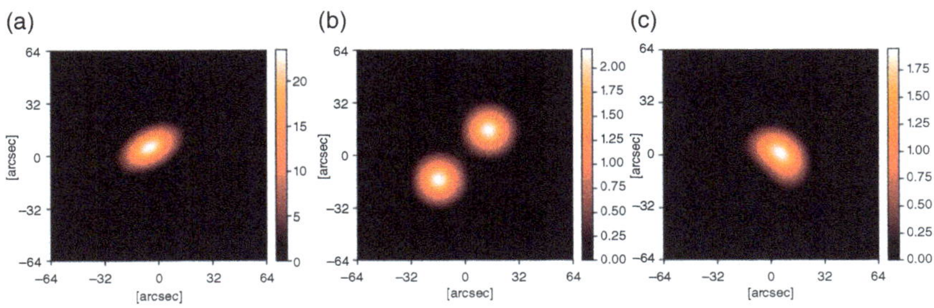

Figure 18.1 (a–c) Examples of source configurations for astrophysical imaging.

19
Conclusions

Kernel-based approximation has undergone a remarkable evolution over the past decades, and the material presented in this book reflects both its classical foundations and its most recent developments. The journey from positive-definite kernels to state-of-the-art variably scaled and data-driven constructions demonstrates how the field has matured into a versatile framework capable of addressing problems across numerical analysis, scientific computing, machine learning, imaging, and beyond.

A central theme of this volume is that kernel methods naturally bridge rigorous functional analysis with practical numerical tools. Reproducing kernel Hilbert spaces provide the analytical backbone, offering an elegant setting in which approximation, stability, and regularity can be quantified. At the same time, the radial basis function perspective enables practitioners to work effectively with scattered, high-dimensional data, free from the constraints of meshes. In Chapters 5, 6, 7, and 8, we have emphasized the delicate interplay between accuracy and numerical stability, particularly in the context of shape-parameter tuning, rational kernels, and reduced-basis selection strategies. These aspects underscore the fact that kernel methods are not only theoretically robust but also computationally adaptable.

The development of VSKs, especially their discontinuous variants (VS-DKs), has been one of the most significant contributions of recent years. These constructions provide a principled way to overcome the classical Gibbs phenomenon and to maintain high accuracy even in the presence of edges and discontinuities. The applications of magnetic particle imaging and astronomical imaging illustrate how theoretical advances can be quickly translated into concrete tools useful in real-world scenarios, including those with noisy, irregular, or incomplete data.

The second half of the book demonstrates the reverse direction of influence: modern machine learning, far from being detached from approximation theory, benefits from kernel-based insights. Support vector machines regularization networks, and local-to-global classifiers exemplify how classical kernels can be enhanced using domain decomposition, partition-of-unity techniques, and reduced-complexity formulations. Furthermore, recent developments in learning optimal shape matrices, data-driven kernel optimization (e.g., 2L-KOGA), and the use of discontinuous neural networks to learn scaling functions reveal a paradigm in which learning and approximation reinforce each other.

While we have aimed to provide a broad introduction and a coherent picture of modern kernel approximation, several essential topics lie outside the scope of this volume. First, we did not treat kernels statistically. Concepts such as Gaussian processes, Bayesian interpretation of kernel regression, covariance kernels derived from stochastic processes, or uncertainty quantification through posterior variances were intentionally omitted, as they belong to a rich and active line of research (Gardner et al., 2018; Paciorek and Schervish, 2003; Wilson and Nickisch, 2015).

Second, although we briefly mentioned collocation and meshfree methods, we did not develop kernel-based PDE solvers, including RBF collocation, symmetric/unsymmetric methods, RBF–FD (radial basis function–finite differences), Hermite-type formulations, and stability issues related to PDE operators. These topics require a dedicated treatment that combines approximation theory with numerical analysis for differential operators (Akers et al., 2021; Erb et al., 2024; Andrew et al., 2023; Tominec et al., 2021, 2025; Reeger, 2024; Shankar et al., 2021).

Lastly, the book does not address kernels on non-Euclidean domains such as manifolds, graphs, and general metric spaces, despite the growing interest in geometric data analysis and graph-based learning (Hangelbroek and Rieger, 2025; Hangelbroek et al., 2011, 2025; Jayasumana et al., 2015).

In conclusion, the objective of this book has been to provide a clear, accessible, and mathematically grounded entry point to the field of kernel approximation, while simultaneously exposing the reader to recent advances introduced by the authors. We hope that these notes will serve both as an introduction for newcomers and as a reference for researchers seeking a structured overview of modern kernel-based approaches. The field continues to evolve rapidly, propelled by the expanding role of data-driven modeling, and we expect kernel methods to remain central in many scientific and engineering applications.

This book is certainly not an alternative to books already well known in the RBF and meshless community, in particular those that inspired us, by G.

E. Fasshauer Fasshauer (2007), G. E. Fasshauer and M.J. McCourt Fasshauer and McCourt (2015), and H. Wendland Wendland (2005). Nevertheless, if these pages have inspired the reader to explore further, develop, or apply kernel techniques to modern problems, then the purpose of this work has been achieved.

References

Adams, H., Emerson, T., Kirby, M., et al. 2017. Persistence images: a stable vector representation of persistent homology. *Journal of Machine Learning Research*, **18**, 1–35.

Adams, R. A., and Fournier, J. 2003. *Sobolev Spaces*. London: Academic Press.

Aggarwal, C. C. 2014. *Data Classification: Algorithms and Applications*. Boca Raton, FL: CRC Press.

Akers, B., Liu, T., and Reeger, J. 2021. A radial basis function finite difference scheme for the Benjamin–Ono equation. *Mathematics*, **9**(1), 65.

Albrecht, K., and Iske, A. 2025. On the convergence of generalized kernel-based interpolation by greedy data selection algorithms. *BIT Numerical Mathematics*, **65**(1).

Albrecht, K., Entzian, J., and Iske, A. 2025. Product kernels are efficient and flexible tools for high-dimensional scattered data interpolation. *Advances in Computational Mathematics*, **51**(2).

Allasia, G., Besenghi, R., Cavoretto, R., and De Rossi, A. 2011. Scattered and track data interpolation using an efficient strip searching procedure. *Applied Mathematics and Computation*, **217**, 5949–5966.

Allavena, S., Piana, M., Benvenuto, F., and Massone, A. M. 2012. An interpolation/extrapolation approach to X-ray imaging of solar flares. *Inverse Problems and Imaging*, **6**, 147.

Alon, N., Ben-David, S., Cesa-Bianchi, N., and Haussler, D. 1997. Scale-sensitive dimensions, uniform convergence, and learnability. *Journal of the ACM*, **44**, 615–631.

Altman, N. S. 1992. An Introduction to kernel and nearest-neighbor nonparametric regression. *The American Statistician*, **46**(3), 175–185.

Andrew, M. J., Bosler, P. A., Kuberry, P. A., and Wright, G. B. 2023. Generalized moving least squares vs. radial basis function finite difference methods for approximating surface derivatives. *Computers & Mathematics with Applications*, **147**, 1–13.

Aronszajn, N. 1950. Theory of reproducing kernels. *Transactions of the American Mathematical Society*, **68**, 337–404.

Arya, S., Mount, D. M., Netanyahu, N. S., Silverman, R., and Wu, A. Y. 1998. An optimal algorithm for approximate nearest neighbor searching in fixed dimensions. *Journal of the ACM*, **45**, 891–923.

Audone, G., Della Santa, F., Perracchione, E., and Pieraccini, S. 2025. A recipe based on Lebesgue functions for learning variably scaled kernels via discontinuous neural networks. *Journal of Computational and Applied Mathematics*, **469**, 116653.

Avesani, S., Kempf, R., Multerer, M., and Wendland, H. 2025. Multiscale scattered data analysis in samplet coordinates. *SIAM Journal on Scientific Computing*, **47**(5), A3038–A3063.

Babuška, I., and Melenk, J. M. 1997. The partition of unity method. *International Journal for Numerical Methods in Engineering*, **40**, 727–758.

Balcan, M.-F. 2011. *Lecture Notes 24–25 of Machine Learning Theory*. Georgia Tech lecture notes.

Bandiziol, C. 2025. *Applications of Persistent Homology: Data Classification and Intrinsic Dimension of Manifold*. PhD thesis, University of Padova, Padova, Italy. Available at `https://hdl.handle.net/11577/3557744`.

Bandiziol, C., and De Marchi, S. 2024. Persistence symmetric kernels for classification: a comparative study. *Symmetry*, **16**(9). Article 1236.

Bandiziol, C., De Marchi, S., and Allegra, M. 2025. On intrinsic dimension of point clouds by a persistent homology approach: computational tips. *Modern Mathematical Models*, **3**(3), 171–184.

Bartlett, P. L., and Mendelson, S. 2002. Rademacher and Gaussian complexities: risk bounds and structural results. *Journal of Machine Learning Research*, **3**, 463–482.

Bayona, V., Moscoso, M., and Kindelan, M. 2012. Gaussian RBF-FD weights and its corresponding local truncation errors. *Engineering Analysis with Boundary Elements*, **36**(9), 1361–1369.

Berg, M. De, Kreveld, M. Van, Overmars, M., and Schwarzkopf, O. 1997. *Computational Geometry*. Berlin: Springer.

Berrut, J.-P., De Marchi, S., Elefante, G., and Marchetti, F. 2020. Treating the Gibbs phenomenon in barycentric rational interpolation and approximation via the S-Gibbs algorithm. *Applied Mathematics Letters*, **103**, 106196.

Bertero, M., and Piana, M. 2006. Inverse problems in biomedical imaging: modeling and methods of solution. Pages 1–33 of: Alfio Quarteroni, Luca Formaggia, and Alessandro Veneziani (eds), *Complex Systems in Biomedicine*. Springer - Milano.

Bhatia, R. 1997. *Matrix Analysis*. New York: Springer-Verlag.

Bhatt, R., and Dhall, A. *Skin Segmentation Dataset*. UCI Machine Learning Repository.

Blanzieri, E., and Melgani, F. 2006. An adaptive SVM nearest neighbor classifier for remotely sensed imagery. Pages 3931–3934 of: *IEEE International Geoscience and Remote Sensing Symposium*. Denver, CO.

Bochner, S. 1933. Monotone Funktionen, Stieltjes Integrale und harmonische Analyse. *Mathematische Annalen*, **108**, 378–410.

Bohn, B., Rieger, C., and Grieber, M. 2019. A representer theorem for deep kernel learning. *Journal of Machine Learning Research*, **20**(1), 2302–2333.

Borgwardt, K. M., Ong, C. S., Schönauer, S., Vishwanathan, S. V., Smola, A. J., and Kriegel, H. P. 2005. Protein function prediction via graph kernels. *Bioinformatics*, **21**(Suppl. 1), i47–i56.

Bos, L., Caliari, M., De Marchi, S., Vianello, M., and Xu, Y. 2006. Bivariate Lagrange interpolation at the Padua points: the generating curve approach. *Journal of Approximation Theory*, **143**, 15–25.

Bos, L., De Marchi, S., and Vianello, M. 2017. Polynomial approximation on Lissajous curves in the d-cube. *Applied Numerical Mathematics*, **116**, 47–56.

Bozzini, M., Lenarduzzi, L., Rossini, M., and Schaback, R. 2015. Interpolation with variably scaled kernels. *IMA Journal of Numerical Analysis*, **35**, 199–219.

Bredensteiner, E. J., and Bennett, K. P. 1999. Multicategory classification by support vector machines. *Computational Optimization and Applications*, **12**, 53–79.

Breimann, L. 2001. Random forests. *Machine Learning*, **45**, 5–32.

Buhmann, M. D. 2003. *Radial Basis Functions: Theory and Implementations*. Cambridge Monographs on Applied and Computational Mathematics. Cambridge: Cambridge University Press.

Buhmann, M. D., De Marchi, S., and Perracchione, E. 2020. Analysis of a new class of rational RBF expansions. *IMA Journal of Numerical Analysis*, **40**(3), 1972–1993.

Caliari, M., De Marchi, S., and Vianello, M. 2005. Bivariate polynomial interpolation on the square at new nodal sets. *Applied Mathematics and Computation*, **165**(2), 261–274.

Campagna, R., Conti, C., and Cuomo, S. 2019. Smoothing exponential-polynomial splines for multiexponential decay data. *Dolomites Research Notes on Approximation*, **12**, 86–100.

Campi, C., Marchetti, F., and Perracchione, E. 2021. Learning via variably scaled kernels. *Advances in Computational Mathematics*, **47**, 51.

Canny, J. F. 1986. A computational approach to edge detection. *IEEE Transactions on Pattern Analysis and Machine Intelligence*, **8**, 34–41.

Carlsson, G. 2009. Topology and data. *Bulletin of the American Mathematical Society (N.S.)*, **46**, 255–308.

Carr, J. C., Fright, W. R., and Beatson, R. K. 1997. Surface interpolation with radial basis functions for medical imaging. *IEEE Transactions on Medical Imaging*, **16**, 96–107.

Carr, J. C., Beatson, R. K., Cherrie, J. B., Mitchell, T. J., Fright, W. R., McCallum, B. C., and Evans, T. R. 2001. Reconstruction and representation of 3D objects with radial basis functions. Pages 67–76 of: *Proceedings of the 28th Annual Conference on Computer Graphics and Interactive Techniques*. New York: ACM Press.

Carrière, M., Cuturi, M., and Oudot, S. 2017. Sliced Wasserstein kernel for persistence diagrams. Pages 664–673 of: *Proceedings of the 34th International Conference on Machine Learning*, vol. 70. JMLR.org.

Carrière, M., Oudot, S. Y., and Ovsjanikov, M. 2015. Stable topological signatures for points on 3D shapes. *Computer Graphics Forum*, **34**(5), 1–12.

Cavoretto, R., and De Rossi, A. 2010. Fast and accurate interpolation of large scattered data sets on the sphere. *Journal of Computational and Applied Mathematics*, **234**, 1505–1521.

Cavoretto, R., and De Rossi, A. 2012. Spherical interpolation using the partition of unity method: an efficient and flexible algorithm. *Applied Mathematics Letters*, **25**, 1251–1256.

Cavoretto, R., and De Rossi, A. 2014. Achieving accuracy and efficiency in spherical modelling of real data. *Mathematical Methods in the Applied Sciences*, **37**, 1449–1459.

Cavoretto, R., and De Rossi, A. 2020. An adaptive LOOCV-based refinement scheme for RBF collocation methods over irregular domains. *Applied Mathematics Letters*, **103**, 106178.

Cavoretto, R., Fasshauer, G. E., and McCourt, M. 2015. An introduction to the Hilbert-Schmidt SVD using iterated Brownian bridge kernels. *Numerical Algorithms*, **68**, 393–422.

Cavoretto, R., De Rossi, A., and Perracchione, E. 2016. Efficient computation of partition of unity interpolants through a block-based searching technique. *Computers & Mathematics with Applications*, **71**, 2568–2584.

Cavoretto, R., De Rossi, A., and Perracchione, E. 2018. Optimal selection of local approximants in RBF-PU Interpolation. *Journal of Scientific Computing*, **74**, 1–22.

Cavoretto, R., De Rossi, A., Lancellotti, S., and Perracchione, E. 2022. Software implementation of the partition of unity method. *Dolomites Research Notes on Approximation*, **15**, 35–46.

Cavoretto, R., De Rossi, A., Lancellotti, S., and Romaniello, F. 2024. Parameter tuning in the radial kernel-based partition of unity method by Bayesian optimization. *Journal of Computational and Applied Mathematics*, **451**, 116108.

Chen, C. S., Karageorghis, A., and Zheng, H. 2020. Improved RBF collocation methods for fourth order boundary value problems. *Communications in Computational Physics*, **27**, 1530–1549.

Chen, X., Golovinskiy, A., and Funkhouser, T. 2009. A Benchmark for 3D Mesh Segmentation. *ACM Transactions on Graphics (Proc. SIGGRAPH)*, **28**(3), 1–12.

Chen, Y. L., and Lai, S. H. 2007. A partition of unity based algorithm for implicit surface reconstruction using belief propagation. Pages 147–152 of: *IEEE International Conference on Shape Modeling and Applications*. Los Alamitos: IEEE.

Chiu, S. N., Ling, L., and McCourt, M. 2020. On variable and random shape Gaussian interpolations. *Applied Mathematics and Computation*, **377**, 125159.

Chung, K. C., and Yao, T. H. 1977. On lattices admitting unique Lagrange interpolations. *SIAM Journal on Numerical Analysis*, **14**, 735–743.

Cohen-Steiner, D., Edelsbrunner, H., and Harer, J. 2007. Stability of persistence diagrams. *Discrete and Computational Geometry* **37**, 103–120.

Collins, M., and Duffy, N. 2001. Convolution kernels for Natural Language. *In Proceedings of the 15th International Conference on Neural Information Processing Systems: Natural and Synthetic (NIPS'01)*, 625–632. Cambridge, MA: MIT Press.

Collobert, R., Bengio, S., and Bengio, Y. 2001. A parallel mixture of SVMs for very large scale problems. Pages 633–640 of: T. Dietterich, S. Becker, and Z. Ghahramani (eds), *NIPS*.

Cortes, C., and Vapnik, V. 1995. Support vector networks. *Machine Learning*, **20**, 273–297.

Cuomo, S., Galletti, A., Giunta, G., and Starace, A. 2013. Surface reconstruction from scattered point via RBF interpolation on GPU. Pages 433–440 of: *Federated Conference on Computer Science and Information Systems*. Los Alamitos: IEEE.

Cuomo, S., Galletti, A., Giunta, G., and Marcellino, L. 2017. Reconstruction of implicit curves and surfaces via RBF interpolation. *Applied Numerical Mathematics*, 157–171.

Curtis, P. C. Jr. 1959. Parameter families and best approximation. *Pacific Journal of Mathematics*, **9**, 1013–1027.

De Marchi, S. *Short Introduction to Topological Data Analysis, Persistent Homology and Applications*. `https://www.math.unipd.it/~demarchi/CorsoTDA/CorsoTDAUppsala.pdf`.

De Marchi, S. 2003. On optimal center locations for radial basis function interpolation: computational aspects. *Rendiconti del Seminario Matematico Universitá Politecnica di Torino*, **61**, 343–358.

De Marchi, S., and Santin, G. 2013. A new stable basis for radial basis function interpolation. *Journal of Computational and Applied Mathematics*, **253**, 1–13.

De Marchi, S., and Santin, G. 2015. Fast computation of orthonormal basis for RBF spaces through Krylov space methods. *BIT Numerical Mathematics*, **55**, 949–966.

De Marchi, S., Schaback, R., and Wendland, H. 2005. Near-optimal data-independent point locations for radial basis function interpolation. *Advances in Computational Mathematics*, **23**, 317–330.

De Marchi, S., Idda, A., and Santin, G. 2016. A Rescaled Method for RBF Approximation. Pages 39–59 of: et al., G. E. Fasshauer (ed.), *Approximation Theory XV: San Antonio 2016*, vol. 201. Springer International Publishing.

De Marchi, S., Erb, W., and Marchetti, F. 2017. Spectral filtering for the reduction of the Gibbs phenomenon for polynomial approximation methods on Lissajous curves with applications in MPI. *Dolomites Research Notes on Approximation*, **10**, 128–137.

De Marchi, S., Martínez, A., and Perracchione, E. 2019. Fast and stable rational RBF-based partition of unity interpolation. *Journal of Scientific Computing*, **79**, 321–344.

De Marchi, S., Erb, W., Francomano, E., Marchetti, F., Perracchione, E., and Poggiali, D. 2020a. Fake nodes approximation for magnetic particle imaging. *2020 IEEE 20th Mediterranean Electrotechnical Conference (MELECON) Proceedings*, 434–438.

De Marchi, S., Marchetti, F., and Perracchione, E. 2020b. Jumping with variably scaled discontinuous kernels (VSDKs). *BIT Numerical Mathematics*, **60**(2), 441–463.

De Marchi, S., Marchetti, F., Perracchione, E., and Poggiali, D. 2020c. Multivariate approximation at fake nodes. *Applied Mathematics and Computation*. accepted.

De Marchi, S., Marchetti, F., Perracchione, E., and Poggiali, D. 2020d. Polynomial interpolation via mapped bases without resampling. *Journal of Computational and Applied Mathematics*, **364**, 112347.

De Marchi, S., Erb, W., Marchetti, F., Perracchione, E., and Rossini, M. 2020e. Shape-driven interpolation with discontinuous kernels: error analysis, edge extraction and applications in MPI. *SIAM Journal on Scientific Computing*, **42**(2), B472–B491.

De Marchi, S., Lot, F., Marchetti, F., and Poggiali, D. 2022. Variably scaled persistence kernels (VSPKs) for persistent homology applications. *Journal of Computational Mathematics and Data Science*, **4**, 100050.

Dekel, O., and Shamir, O. 2012. There's a hole in my data space: Piecewise predictors for heterogeneous learning problems. *Journal of Machine Learning Research*, **22**, 291–298.

Della Santa, F., and Pieraccini, S. 2023. Discontinuous neural networks and discontinuity learning. *Journal of Computational and Applied Mathematics*, **419**, 114678.

Deng, X., Coles, W., Hobbs, G., Keith, M., Manchester, R., Shannon, R., and Zheng, J. 2012. Optimal interpolation and prediction in pulsar timing. *Monthly Notices of the Royal Astronomical Society*, **424**, 244–251.

Diederichs, B., and Iske, A. 2019. Improved estimates for condition numbers of radial basis function interpolation matrices. *Journal of Approximation Theory*, **238**, 38–51.

Ding, S., Zhang, X., An, Y., and Xue, Y. 2017. Weighted linear loss multiple birth support vector machine based on information granulation for multi-class classification. *Pattern Recognition*, **67**, 32.

Donini, M., and Aiolli, F. 2017. Learning deep kernels in the space of dot product polynomials. *Machine Learning*, **106**, 1245–1269.

Drake, K. P., Fuselier, E. J., and Wright, G. B. 2022. Implicit surface reconstruction with a curl-free radial basis function partition of unity method. *SIAM Journal on Scientific Computing*, **44**(5), A3018–A3040.

Driscoll, T. A., and Fornberg, B. 2002. Interpolation in the limit of increasingly flat radial basis functions. *Computers & Mathematics with Applications*, **43**, 413–422.

Dutta, S., Farthing, M. W., Perracchione, E., Savant, G., and Putti, M. 2021. A greedy non-intrusive reduced order model for shallow water equations. *Journal of Computational Physics*, **439**, 110378.

Easton, R. L. Jr. 2010. *Fourier Methods in Imaging*. John Wiley & Sons.

Edelsbrunner, H., Letscher, D., and Zomorodian, A. 2000. Topological persistence and simplification. *Proceedings of the 41st Annual Symposium on Foundations of Computer Science*, 454–463, IEEE.

Engl, H. W., Hanke, M., and Neubauer, A. 1996. *Regularization of Inverse Problems*. Vol. 375. Dordrecht: Kluwer Academic Publishers.

Erb, W. 2016. Bivariate Lagrange interpolation at the node points of Lissajous curves - the degenerate case. *Applied Mathematics and Computation*, **289**, 409–425.

Erb, W., Kaethner, C., Dencker, P., and Ahlborg, M. 2015. A survey on bivariate Lagrange interpolation on Lissajous nodes. *Dolomites Research Notes on Approximation*, **8**, 23–36.

Erb, W., Kaethner, C., Ahlborg, M., and Buzug, T. M. 2016. Bivariate Lagrange interpolation at the node points of non-degenerate Lissajous nodes. *Numerische Mathematik*, **133**(1), 685–705.

Erb, W., Hangelbroek, T., Narcowich, F.J, Rieger, C., and Ward, J.D. 2024. Highly localized RBF Lagrange functions for finite difference methods on spheres. *BIT Numerical Mathematics*, **64**(2), 16.

Fasshauer, G. E. 2007. *Meshfree Approximation Methods with MATLAB*. Singapore: World Scientific.

Fasshauer, G. E., and McCourt, M. J. 2015. *Kernel-based Approximation Methods Using MATLAB*. Singapore: World Scientific.

Fasshauer, G. E., and Zhang, J. G. 2007. On choosing "optimal" shape parameters for RBF approximation. *Numerical Algorithms*, **45**(1–4), 345–368.

Fletcher, R. 2000. *Practical Methods of Optimization*. 2nd edn. New York: Wiley.

Fomenko, A. 1994. *Visual Geometry and Topology*. Berlin: Springer-Verlag. translated from the Russian by Marianna V. Tsaplina.

Fornberg, B., and Flyer, N. 2008. The Gibbs phenomenon for radial basis functions. In: *The Gibbs Phenomenon in Various Representations and Applications*. Potsdam, NY: Sampling Publishing.

Fornberg, B., and Piret, C. 2008. On choosing a radial basis function and a shape parameter when solving a convective PDE on a sphere. *Journal of Computational Physics*, **227**(5), 2758–2780.

Fornberg, B., and Wright, G. 2004. Stable computation of multiquadric interpolants for all values of the shape parameter. *Computers & Mathematics with Applications*, **48**, 853–867.

Fornberg, B., Larsson, E., and Flyer, N. 2011. Stable Computations with Gaussian Radial Basis Functions. *SIAM Journal on Scientific Computing*, **33**(2), 869–892.

Gamper, U., Boesiger, P., and Kozerke, S. 2008. Compressed sensing in dynamic MRI. *Magnetic Resonance in Medicine*, **59**, 365–373.

Gardner, J., Pleiss, G., Wu, R., Weinberger, K., and Wilson, A. 2018. Product kernel interpolation for scalable Gaussian processes. Pages 1407–1416 of: *International Conference on Artificial Intelligence and Statistics*. PMLR.

Gary, D. E. 2023. New insights from imaging spectroscopy of solar radio emission. *Annual Review of Astronomy and Astrophysics*, **61**, 427–472.

Gleich, B., and Weizenecker, J. 2005. Tomographic imaging using the nonlinear response of magnetic particles. *Nature*, **435**, 1214–1217.

Golub, G. H., Heath, M., and Wahba, G. 1979. Generalized cross-validation as a method for choosing a good ridge parameter. *Technometrics*, **21**(2), 215–223.

Goodfellow, I., Bengio, Y., and Courville, A. 2016. *Deep Learning*. MIT Press. Available at `http://www.deeplearningbook.org`.

Graeser, M., Thieben, F., Szwargulski, P., et al. 2019. Human-sized magnetic particle imaging for brain applications. *Nature Communications*, **10**, 1936.

Guastavino, S., and Benvenuto, F. 2020. Convergence rates of spectral regularization methods: A comparison between ill-posed inverse problems and statistical kernel learning. *SIAM Journal on Numerical Analysis*, **58**(6), 3504–3529.

Guillemard, M., and Iske, A. 2017. Interactions between kernels, frames, and persistent homology. Pages 861–888 of: *Recent Applications of Harmonic Analysis to Function Spaces, Differential Equations, and Data Science: Novel Methods in Harmonic Analysis*, vol. 2. Cham, Switzerland: Springer International Publishing.

Haar, A. 1918. Die Minkowskische Geometrie und die Annäherung an stetige Funktionen. *Mathematische Annalen*, **18**, 294–311.

Hangelbroek, T., and Rieger, C. 2025. Kernel multigrid on manifolds. *Journal of Complexity*, **86**, 101900.

Hangelbroek, T., Narcowich, F. J., Sun, X., and Ward, J. D. 2011. Kernel Approximation on Manifolds II: The L_∞ Norm of the L_2 Projector. *SIAM Journal on Mathematical Analysis*, **43**(2), 662–684.

Hangelbroek, T., Rieger, C., and Wright, G. B. 2025. Spectral stability and perturbation results for kernel differentiation matrices on the sphere. *Numerische Mathematik*, 1–34.

Harris, D., and Harris, S. 2013. *Digital Design and Computer Architecture*. 2nd edn. Boston: Morgan Kaufmann.

Heryudono, A., and Raessi, M. 2023. Adaptive partition of unity interpolation method with moving patches. *Mathematics and Computers in Simulation*, **210**, 49–65.

Ho, T. K. 1995. Random decision forests. Page 278 of: *Proceedings of the Third International Conference on Document Analysis and Recognition*, vol. 1.

Hofmann, T., Schölkopf, B., and Smola, A. J. 2008. Kernel methods in machine learning. *Annals of Statistics*, **36**(3), 1171–1220.

Hoppe, H. 1994. *Surface Reconstruction from Unorganized Points*. Ph.D. thesis, University of Washington, Seattle, WA.

Hoppe, H., Derose, T., Duchamp, T., McDonald, J., and Stuetzle, W. 1992. Surface reconstruction from unorganized points. Pages 71–78 of: *Proceedings of the 19th Annual Conference on Computer Graphics and Interactive Techniques*, vol. 26. New York, NY: ACM Press.

Horak, D., Maletić, S., and Rajković, M. 2009. Persistent homology of complex networks. *Journal of Statistical Mechanics: Theory and Experiment*, P03034.

Horn, R. A., and Johnson, C. R. 2013. *Matrix Analysis*. 2nd edn. Cambridge University Press.

Horn, R. A., and Zhang, F. 2010. Bounds on the spectral radius of a Hadamard product of nonnegative or positive semidefinite matrices. *Electronic Journal of Linear Algebra*, **20**, 90–94.

Huang, H.-Y., and Lin, C.-J. 2016. Linear and kernel classification: When to use which? Pages 216–224 of: *SIAM International Conference on Data Mining*. Philadelphia, PA.

Jakobsson, S., Andersson, B., and Edelvik, F. 2009. Rational radial basis function interpolation with applications to antenna design. *Journal of Computational and Applied Mathematics*, **233**, 889–904.

Jayasumana, S., Hartley, R., Salzmann, M., Li, H., and Harandi, M. 2015. Kernel methods on Riemannian manifolds with Gaussian RBF kernels. *IEEE Transactions on Pattern Analysis and Machine Intelligence*, **37**(12), 2464–2477.

Joachims, T., and Yu, C. N. J. 2009. Sparse kernel SVMs via cutting-plane training. Pages 179–193 of: *Machine Learning*.

Jose, C., Goyal, P., Aggrwal, P., and Varma, M. 2013. Local deep kernel learning for efficient non-linear SVM prediction. Pages 486–494 of: *Proceedings of the 30th International Conference on Machine Learning*. Atlanta, GA.

Kääriäinen, M. 2004. *Relating the Rademacher and VC Bounds*. Tech. rept. Series of Publications C, Report C-2004-57, Helsinki.

Kaethner, C., Erb, W., Ahlborg, M., Szwargulski, P., Knopp, T., and Buzug, T. M. 2016. Non-equispaced system matrix acquisition for magnetic particle imaging based on Lissajous node points. *IEEE Transactions on Medical Imaging*, **35**(11), 2476–2485.

Karoui, N. El. 2010. The spectrum of kernel random matrices. *Annals of Statistics*, **38**, 1–50.

Kearns, M. J., and Schapire, R. E. 1994. Efficient distribution-free learning of probabilistic concepts. *Journal of Computer and System Sciences*, **48**(3), 464–497.

Kempf, R., and Wendland, H. 2023. High-dimensional approximation with kernel-based multilevel methods on sparse grids. *Numerische Mathematik*, **154**(3–4), 485–519.

Khare, K., Butola, M., and Rajora, S. 2015. *Fourier Optics and Computational Imaging*. John Wiley & Sons: Chichester, UK.

Kidger, P., and Lyons, T. 2010. Universal Approximation with Deep Narrow Networks. Pages 2306–2327 of: *Proceedings of Machine Learning Research*, vol. 125.

Kingma, D. P., and Ba, J. 2015. Adam: A Method for Stochastic Optimization. In: *Proceedings of the 3rd International Conference on Learning Representations (ICLR)*.

Knopp, T., and Buzug, T. M. 2012. *Magnetic Particle Imaging*. Berlin Heidelberg: Springer-Verlag.

Knopp, T., Biederer, S., Sattel, T., Weizenecker, J., Gleich, B., Borgert, J., and Buzug, T. M. 2009. Trajectory analysis for magnetic particle imaging. *Physics in Medicine and Biology*, **54**, 385–397.

Kohavi, R. 1996. Scaling up the accuracy of Naive-Bayes classifiers: A decision-tree hybrid. In: *Proceedings of the Second International Conference on Knowledge Discovery and Data Mining*. Portland, OR.

Koltchinskii, V., and Panchenko, D. 2000. Rademacher processes and bounding the risk of function learning. Pages 443–459 of: *High Dimensional Probability II*.

Kriege, N. M., Johansson, F. D., and Morris, C. 2020. A survey on graph kernels. *Applied Network Science*, **5**(6), 385–397.

Kung, S. Y. 2014. *Kernel Methods and Machine Learning*. Cambridge University Press.

Kusano, G., Fukumizu, K., and Hiraoka, Y. 2017. Kernel method for persistence diagrams via kernel embedding and weight factor. *Journal of Machine Learning Research*, **18**, Paper No. 189.

Lagendijk, R. L., Biemond, J., and Boekee, D. E. 1988. Regularized iterative image restoration with ringing reduction. *IEEE Transactions on Signal Processing*, **36**, 1874–1884.

LaMontagne, P. J., Benzinger, T. L. S., Morris, J. C., Keefe, S., Hornbeck, R., Xiong, C., Grant, E., Hassenstab, J., Moulder, K., Vlassenko, A. G., Raichle, M. E., Cruchaga, C., and Marcus, D. 2019. OASIS-3: Longitudinal Neuroimaging, Clinical, and Cognitive Dataset for Normal Aging and Alzheimer Disease. *medRxiv*.

Lampert, C. H. 2009. Kernel methods in computer vision. Pages 193–285 of: *Foundations and Trends in Computer Vision*, vol. 4. Hanover, MA: Now Publishers Inc.

Larkman, D. J., and Nunes, R. G. 2007. *Parallel Magnetic Resonance Imaging*. IOP Publishing/Physics in Medicine & Biology (review). Physics in Medicine & Biology, 52 (2007), R15.

Larsson, E., Lehto, E., Heryudono, A. R. H., and Fornberg, B. 2013. Stable computation of differentiation matrices and scattered node stencils based on Gaussian radial basis functions. *SIAM Journal on Scientific Computing*, **33**, 869–892.

Le, T., and Yamada, M. 2018. Persistence fisher kernel: A Riemannian manifold kernel for persistence diagrams. Pages 10028–10039 of: *NIPS'18*. Curran Associates Inc.

Lin, R. P., Dennis, B. R., Hurford, G. J., Smith, D. M., Zehnder, A., Harvey, P. R., Curtis, D. W., et al. 2002. The Reuven Ramaty high-energy solar spectroscopic imager (RHESSI). *Solar Physics*, **210**, 3–32.

Ling, L., and Marchetti, F. 2022. A stochastic extended Rippa's algorithm for LpOCV. *Applied Mathematics Letters*, **129**, 107955.

Lingala, S. G., Hu, Y., DiBella, E., and Jacob, M. 2011. Accelerated dynamic MRI exploiting sparsity and low-rank structure: kt slr. *IEEE Transactions on Medical Imaging*, **30**, 1042–1054.

Lyche, T. 1992. Knot Removal for Spline Curves and Surfaces. Pages 207–226 of: E. W. Cheney, C. K. Chui and L. L. Schumaker (eds.), *Approximation Theory*. Boston, MA: Academic Press.

Mairhuber, J. C. 1956. On Haar's Theorem concerning Chebychev approximation problems having unique solutions. *Proceedings of the American Mathematical Society*, **7**, 609–615.

Maji, S., Berg, A. C., and Malik, J. 2013. Efficient classification for additive kernel SVMs. *IEEE Transactions on Pattern Analysis and Machine Intelligence*, **35**(1), 66–77.

Marchetti, F. 2021. The extension of Rippa's algorithm beyond LOOCV. *Applied Mathematics Letters*, **120**, 107262.

Marchetti, F. 2024. A fast surrogate cross validation algorithm for meshfree RBF collocation approaches. *Applied Mathematics and Computation*, **481**, 128943.

Marchetti, F., and Perracchione, E. 2022a. Efficient reduced basis algorithm (ERBA) for kernel-based approximation. *Journal of Scientific Computing*, **91**, 41.

Marchetti, F., and Perracchione, E. 2022b. Local-to-global support vector machines (LGSVMs). *Pattern Recognition*, **132**, 108920.

Marchetti, F., Martino, F. De, Shamseddin, M., De Marchi, S., and Brisken, C. 2020. Variably scaled kernels improve classification of hormonally-treated patient-derived xenografts. Pages 1–6 of: *2020 IEEE Conference on Evolving and Adaptive Intelligent Systems (EAIS)*. Bari (Italy), May 27-29, IEEE.

Maron, M. E. 1961. Automatic indexing: An experimental inquiry. *Journal of the ACM*, **8**, 404–417.

Massa, P., Schwartz, R., Tolbert, A. K., Massone, A. M., Dennis, B. R., Piana, M., and Benvenuto, F. 2020. MEM_GE: A new maximum entropy method for image reconstruction from solar X-ray visibilities. *Astrophysical Journal*, **894**, 46.

Massone, A. M., Piana, M., and Prato, M. 2008. Regularized solution of the solar bremsstrahlung inverse problem: Model dependence and implementation issues. *Inverse Problems in Science and Engineering*, **16**, 523–545.

Matérn, B. 1986. *Spatial Variation*. Lecture Notes in Statistics, vol. 36. Berlin: Springer-Verlag.

McLain, D. H. 1976. Two dimensional interpolation from random data. *Computer Journal*, **19**, 178–181.

Melenk, J. M., and Babuška, I. 1996. The partition of unity finite element method: Basic theory and applications. *Computer Methods in Applied Mechanics and Engineering*, **139**, 289–314.

Menon, A. K. 2009. *Large-scale Support Vector Machines: Algorithms and theory*. Technical report. University of California San Diego.

Mercer, J. 1909. Functions of positive and negative type and their connection with the theory of integral equations. *Philosophical Transactions of the Royal Society*, **209**, 415–446.

Meyer, C. D. 2000. *Matrix Analysis and Applied Linear Algebra*. Philadelphia: SIAM.

Micchelli, C. A. 1986. Interpolation of scattered data: Distance matrices and conditionally positive definite functions. *Constructive Approximation*, **2**, 11–22.

Mitchell, T. 1997. *Machine Learning*. New York: McGraw-Hill.

Morse, B., Yoo, T. S., Rheingans, P., Chen, D. T., and Subramanian, K. R. 2001. Interpolating implicit surfaces from scattered surface data using compactly supported radial basis functions. Pages 89–98 of: *IEEE International Conference on Shape Modeling and Applications*. IEEE.

Murphy, K. P. 2012. *Machine Learning: A Probabilistic Perspective*. Cambridge, MA: MIT Press.

Narcowich, F. J., Ward, J. D., and Wendland, H. 2005. Sobolev bounds on functions with scattered zeros, with applications to radial basis function surface fitting. *Mathematics of Computation*, **74**(250), 743–763.

Neumaier, A. 1998. Solving ill-conditioned and singular linear systems: A tutorial on regularization. *SIAM Review*, **40**(3), 636–666.

Ohtake, Y., Belyaev, A., and Seidel, H. P. 2006. Sparse surface reconstruction with adaptive partition of unity and radial basis functions. *Computer Graphics Forum/Graphical Models*, **68**, 15–24.

Olson, T. 2017. *Applied Fourier Analysis*.

Open MPI data website. *OpenMPIData*. `https://magneticparticleimaging.github.io/OpenMPIData.jl/latest/index.html`.

Orr, M. J. L. 1996. *Introduction to Radial Basis Function Networks*. Technical report. University of Edinburgh, Centre for Cognitive Sciences.

Otter, N., Porter, M. A., Tillmann, U., et al. 2017. A roadmap for the computation of persistent homology. *EPJ Data Science*, **6**, 17.

Pachauri, D., Hinrichs, C., Chung, M. K., Johnson, S. C., and Singh, V. 2011. Topology-based kernels with application to inference problems in Alzheimer's disease. *IEEE Transactions on Medical Imaging*, **30**(10), 1760–1770.

Paciorek, C., and Schervish, M. 2003. Nonstationary covariance functions for Gaussian process regression. *Advances in Neural Information Processing Systems*, **16**.

Pallaschke, D., and Rolewicz, S. 1997. Foundations of mathematical optimization. In: *Foundations of Mathematical Optimization*. Dordrecht: Springer.

Pedegrosa, F. et al. 2011. Scikit-learn: Machine learning in Python. *The Journal of Machine Learning Research*, **12**, 2825–2830.

Pelletier, C., Webb, G. I., and Petitjean, F. 2019. Temporal convolutional neural network for the classification of satellite image time series. *Remote Sensing*, **11**(5), 523.

Perracchione, E. 2017. *RBF-Based Partition of Unity Method: Theory, Algorithms and Applications*. PhD thesis, PhD Thesis, University of Torino.

Perracchione, E. 2018. Rational RBF-based partition of unity method for efficiently and accurately approximating 3D objects. *Computers & Applied Mathematics*, **37**, 4633–4648.

Perracchione, E., Massone, A. M., and Piana, M. 2021. Feature augmentation for the inversion of the Fourier transform with limited data. *Inverse Problems*, **37**, 105001.

Perracchione, E., Camattari, F., Volpara, A., Massa, P., Massone, A. M., and Piana, M. 2023. Unbiased CLEAN for STIX in Solar Orbiter. *Astrophysical Journal Supplement Series*, **268**.

Perron, O. 1907. Zur Theorie der Matrices. *Mathematische Annalen*, **64**, 248–263.

Piana, M., Massone, A. M., Hurford, G., Prato, M., Emslie, A. G., Kontar, E. P., and Schwartz, R. A. 2007. Electron flux spectral imaging of solar flares through regularized analysis of hard x-ray source visibilities. *Astrophysical Journal*, **665**, 846–855.

Piana, M., Emslie, A. G., Massone, A. M., and Dennis, B. R. 2022. *Hard X-ray Imaging of Solar Flares*. Springer.

Prévost, C., Borsoi, R. A., Usevich, K., Brie, D., Bermudez, J. C., and Richard, C. 2022. Hyperspectral super-resolution accounting for spectral variability: Coupled tensor ll1-based recovery and blind unmixing of the unknown super-resolution image. *SIAM Journal on Imaging Sciences*, **15**, 110–138.

Ravishankar, S., and Bresler, Y. 2015. Efficient blind compressed sensing using sparsifying transforms with convergence guarantees and application to magnetic resonance imaging. *SIAM Journal on Imaging Sciences*, **8**, 2519–2557.

Reeger, J. A. 2024. Adaptivity in local kernel based methods for approximating the action of linear operators. *SIAM Journal on Scientific Computing*, **46**, A2683–A2708.

Reininghaus, J., Huber, S., Bauer, U., and Kwitt, R. 2015. A stable multi-scale kernel for topological machine learning. Pages 4741–4748 of: *2015 IEEE Conference on Computer Vision and Pattern Recognition (CVPR)*. Boston, MA: IEEE.

Ren, J., Lee, S. D., Chen, X., et al. 2009. Naive Bayes classification of uncertain data. Pages 944–949 of: *Proceedings of the 9th IEEE International Conference Data Mining (ICDM)*.

Rieger, C., and Wendland, H. 2024. On the approximability and curse of dimensionality of certain classes of high-dimensional functions. *SIAM Journal on Numerical Analysis*, **62**(2), 842–871.

Rieger, C., and Zwicknagl, B. 2010. Sampling inequalities for infinitely smooth functions, with applications to interpolation and machine learning. *Advances in Computational Mathematics*, **32**(1), 103–129.

Riesz, F. 1909. Sur le oprérations fonctionnelles linéaires. *Comptes Rendus de l'Académie des Sciences*, **149**, 974–977.

Rippa, S. 1999. An algorithm for selecting a good value for the parameter c in radial basis function interpolation. *Advances in Computational Mathematics*, **11**, 193–210.

Rokach, L., and Maimon, O. 2014. *Data Mining With Decision Trees: Theory and Applications*. River Edge, NJ: World Scientific Publishing Co., Inc.

Romani, L., Rossini, M., and Schenone, D. 2018. Edge detection methods based on RBF interpolation. *Journal of Computational and Applied Mathematics*, **349**, 532–547.

Rossini, M. 2018. Interpolating functions with gradient discontinuities via variably scaled kernels. *Dolomites Research Notes on Approximation*, **11**, 3–14.

Safdari-Vaighani, A., Heryudono, A., and Larsson, E. 2015. A radial basis function partition of unity collocation method for convection-diffusion equations arising in financial applications. *Journal of Scientific Computing*, **64**, 341–367.

Santin, G., and Haasdonk, B. 2017. Convergence rate of the data-independent P-greedy algorithm in kernel-based approximation. *Dolomites Research Notes on Approximation*, **10**, 68–78.

Santin, G., and Haasdonk, B. 2021. *Kernel Methods for Surrogate Modeling*. Vol. 1. De Gruyter.

Sarra, S. A., and Bay, Y. 2018. A rational radial basis function method for accurately resolving discontinuities and steep gradients. *Applied Numerical Mathematics*, **130**, 131–142.

Schaback, R. 1994. Lower bound for norms of inverses of interpolation matrices for radial basis functions. *Journal of Approximation Theory*, **79**, 287–306.

Schaback, R. 1995. Error estimates and condition numbers for radial basis function interpolation. *Advances in Computational Mathematics*, **3**, 251–264.

Schaback, R. 2000. A unified theory of radial basis functions: Native Hilbert spaces for radial basis functions II. *Journal of Computational and Applied Mathematics*, **121**, 165–177.

Schaback, R., and Wendland, H. 2006. Kernel techniques: from machine learning to meshless methods. Pages 543–639 of: *Acta Numerica*, vol. 15.

Schaback, R., and Wu, Z. 1996. Operators on radial functions. *Journal of Computational and Applied Mathematics*, **73**, 257–270.

Schoenberg, I. J. 1938. Metric spaces and completely monotone functions. *Annals of Mathematics*, **39**, 811–841.

Schölkopf, B., and Smola, A. J. 2002. *Learning with Kernels: Support Vector Machines, Regularization, Optimization, and Beyond*. Cambridge, MA: MIT Press.

Schur, J. 1911. Bemerkungen zur Theorie der beschränkten Bilinearformen mit unendlich vielen Veränderlichen. *Journal für die reine und angewandte Mathematik*, **140**, 1–28.

Schwartz, R.A., Csillaghy, A., Tolbert, A.K., Hurford, G.J., McTiernan, J., and Zarro, D. 2002. RHESSI Data Analysis Software: Rationale and Methods. *Solar Physics*, **210**, 165–191.

Shankar, V., Wright, G. B., and Fogelson, A. L. 2021. An efficient high-order meshless method for advection-diffusion equations on time-varying irregular domains. *Journal of Computational Physics*, **445**, 110633.

Shapiro, J. H. 2016. *A Fixed-Point Farrago*. Springer International Publishing.

Sharifi, M., Fathy, M., and Mahmoudi, M. T. 2002. A Classified and Comparative Study of Edge Detection Algorithms. Pages 117–120 of: *Proceedings of International Conference on Information Technology: Coding and Computing*.

Shawe-Taylor, J., and Cristianini, N. 2004. Kernel Methods for Pattern Analysis. In: *Kernel Methods for Pattern Analysis*. Cambridge University Press.

Shawe-Taylor, J., and Sun, S. 2014. Kernel Methods and Support Vector Machines. Pages 857–881 of: *Academic Press Library in Signal Processing, Volume 1*. Elsevier.

Shcherbakov, V., and Larsson, E. 2016. Radial basis function partition of unity methods for pricing vanilla basket options. *Computers & Mathematics with Applications*, **71**, 185–200.

Shepard, D. 1968. A two-dimensional interpolation function for irregularly spaced data. Page 517 of: *Proceedings of 23rd National Conference*. Association for Computing Machinery. New York, NY.

Shepp, L. A., and Logan, B. F. 1974. The Fourier reconstruction of a head section. *IEEE Transactions on Nuclear Science*, **3**, 21–43.

Smola, A. J., and Schölkopf, B. 2004. A tutorial on support vector regression. *Statistics and Computing*, **14**, 199–222.

Solomon, C. J., and Breckon, T. P. 2010. *Fundamentals of Digital Image Processing: A Practical Approach with Examples in Matlab*. Wiley-Blackwell.

Sourbron, S. P., and Buckley, D. L. 2013. Classic models for dynamic contrast-enhanced MRI. *NMR in Biomedicine*, **26**, 1004–1014.

Stura, I., Gabriele, D., and Guiot, C. 2016. A simple PSA-based computational approach predicts the timing of cancer relapse in prostatectomized patients. *Cancer Research*, **76**, 4941–4947.

Su, Y., and Ling, L. 2025. Greedy trial subspace selection in meshfree time-stepping scheme with applications in coupled bulk-surface pattern formations. *Mathematics and Computers in Simulation*, **228**, 498–513.

Sun, J., and Wang, W. 2026. Optimizing shape parameters in RBF methods: A systematic review of techniques, applications, and computational challenges. *Computer Science Review*, **59**, 100842.

Thorsten, J. 1998. Text categorization with Support Vector Machines: Learning with many relevant features. Pages 137–148 of: *Machine Learning: ECML-98*. Lecture Notes in Computer Science (Lecture Notes in Artificial Intelligence), Vol. 1398, Berlin: Springer.

Tibshirani, R., Hastie, T., Narasimhan, B., and Chu, G. 2002. Diagnosis of multiple cancer types by shrunken centroids of gene expression. *Proceedings of the National Academy of Sciences of the United States of America*, **99**(10), 6567–6572.

Tominec, I., Larsson, E., and Heryudono, A. 2021. A least squares radial basis function finite difference method with improved stability properties. *SIAM Journal on Scientific Computing*, **43**(2), A1441–A1471.

Tominec, I., Nazarov, M., and Larsson, E. 2025. Stability estimates for radial basis function methods applied to linear scalar conservation laws. *Applied Mathematics and Computation*, **485**, 129020.

Trahan, C. J., and Wyatt, R. E. 2003. Radial basis function interpolation in the quantum trajectory method: optimization of the multi-quadric shape parameter. *Journal of Computational Physics*, **185**(1), 27–49.

Tsao, J., and Kozerke, S. 2012. MRI temporal acceleration techniques. *Journal of Magnetic Resonance Imaging*, **36**, 543–560.

Turk, G., and O'Brien, J. F. 2002. Modelling with implicit surfaces that interpolate. *ACM Transactions on Graphics*, **21**, 855–873.

van Rijsbergen, C. J. 1979. *Information Retrival*. Butterworth.

Vapnik, V. N. 1982. *Estimation of Dependences Based on Empirical Data*. Berlin: Springer-Verlag.

Vapnik, V. N. 1995. *The Nature of Statistical Learning Theory*. New York: Springer.

Vapnik, V. N. 1998. *Statistical Learning Theory*. Wiley.

Vapnik, V. N., and Chervonenkis, A. Ya. 1971. On the Uniform Convergence of Relative Frequencies of Events to Their Probabilities. *Theory of Probability and its Applications*, **16**(2), 264–280.

Virtanen, P., Gommers, R., et al., Oliphant, and SciPy 1.0 Contributors. 2020. SciPy 1.0: Fundamental algorithms for scientific computing in Python. *Nature Methods*, **17**, 261–272.

Vishwanathan, S. V. N., Schraudolph, Nicol N., Kondor, Risi, and Borgwardt, Karsten M. 2010. Graph kernels. Pages 1201–1242 of: *Kernel Methods*, vol. 11.

Volpara, A., Lupoli, A., Filbir, F., Perracchione, E., Massone, A. M., and Piana, M. 2025. RIS: Regularized imaging spectroscopy for STIX on board Solar Orbiter. *Astronomy & Astrophysics*, **694**, A47.

Wang, X., Chung, F., and Wang, S. 2010. On minimum class locality preserving variance support vector machine. *Pattern Recognition*, **43**(8), 2753–2762.

Wang, Z., Bovik, A. C., Sheikh, H. R., and Simoncelli, E. P. 2004. Image quality assessment: from error visibility to structural similarity. *IEEE Transactions on Image Processing*, **13**(4), 600–612.

Wendland, H. 1995. Piecewise polynomial, positive definite and compactly supported radial functions of minimal degree. *Advances in Computational Mathematics*, **4**, 389–396.

Wendland, H. 2002. Fast evaluation of radial basis functions: Methods based on partition of unity. Pages 473–483 of: *Approximation Theory X: Wavelets, Splines, and Applications*. Nashville: Vanderbilt University Press.

Wendland, H. 2005. *Scattered Data Approximation*. Cambridge Monographs on Applied and Computational Mathematics, vol. 17. Cambridge: Cambridge University Press.

Wendland, H., and Rieger, C. 2005. Approximate interpolation with applications to selecting smoothing parameters. *Numerische Mathematik*, **101**(4), 729–748.

Wenzel, T., Santin, G., and Haasdonk, B. 2021. A novel class of stabilized greedy kernel approximation algorithms: Convergence, stability and uniform point distribution. *Journal of Approximation Theory*, **262**, 105508.

Wenzel, T., Santin, G., and Haasdonk, B. 2023. Analysis of target data-dependent greedy kernel algorithms: Convergence rates for f-, $f \cdot P$- and f/P-Greedy. *Constructive Approximation*, **57**, 45–74.

Wenzel, T., Marchetti, F., and Perracchione, E. 2024a. Data-driven kernel designs for optimized greedy schemes: A machine learning perspective. *SIAM Journal on Scientific Computing*, **46**(1), C101–C126.

Wenzel, T., Santin, G., and Haasdonk, B. 2024b. Stability of convergence rates: Kernel interpolation on non-Lipschitz domains. *IMA Journal of Numerical Analysis*. drae014.

Wenzel, T., Santin, G., and Haasdonk, B. 2025. Stability of convergence rates: Kernel interpolation on non-Lipschitz domains. *IMA Journal of Numerical Analysis*, **45**, 777–798.

Wenzel, Tizian, Santin, Gabriele, and Haasdonk, Bernard. 2026. Analysis of structured deep kernel networks. *Journal of Computational and Applied Mathematics*, **476**.

Wilson, A., and Nickisch, H. 2015. Kernel interpolation for scalable structured Gaussian processes (KISS-GP). Pages 1775–1784 of: *International Conference on Machine Learning*. PMLR.

Wirtz, D., and Haasdonk, B. 2013. A vectorial kernel orthogonal greedy algorithm. *Dolomites Research Notes on Approximation*, **6**, 83–92.

Wirtz, D., Karajan, N., and Haasdonk, B. 2014. Surrogate modeling of multiscale models using kernel methods. *International Journal for Numerical Methods in Engineering*, **101**, 1–28.

Wu, L. C., Zhang, Y., Steinberg, G., et al. 2019. A review of magnetic particle imaging and perspectives on neuroimaging. *American Journal of Neuroradiology*.

Wu, Z., and Schaback, R. 1993. Local error estimates for radial basis function interpolation of scattered data. *IMA Journal of Numerical Analysis*, **13**, 13–27.

Yang, F.L., Yan, L., and Ling, L. 2018. Doubly stochastic radial basis function methods. *Journal of Computational Physics*, **363**, 87–97.

Yao, G., Duo, J., Chen, C. S., and Shen, L. H. 2015. Implicit local radial basis function interpolations based on function values. *Applied Mathematics and Computation*, **265**, 91–102.

Yu, E. Y., Bishop, M., Zheng, B., et al. 2017. Magnetic particle imaging: a novel in vivo imaging platform for cancer detection. *Nano Letters*, **17**, 1648–1654.

Zhang, C., and Ma, Y. 2012. *Ensemble Machine Learning*. Boston, MA: Springer.

Zhang, H., Cao, L., and Gao, S. 2014. A locality correlation preserving support vector machine. *Pattern Recognition*, **47**(9), 3168–3178.

Zhang, J., Su, R., Fu, Q., Ren, W., Heide, F., and Nie, Y. 2022. A survey on computational spectral reconstruction methods from rgb to hyperspectral imaging. *Scientific Reports*, **12**, 11905.

Zhu, S., and Wathen, A. J. 2015. Convexity and solvability for compactly supported radial basis functions with different shapes. *Journal of Scientific Computing*, **63**, 862–884.

Index

For EU product safety concerns, contact us at Calle de José Abascal, 56–1°,
28003 Madrid, Spain or eugpsr@cambridge.org.

www.ingramcontent.com/pod-product-compliance
Ingram Content Group UK Ltd.
Pitfield, Milton Keynes, MK11 3LW, UK
UKHW020016280926
476049UK00002B/12